Reviews of
Environmental Contamination
and Toxicology

VOLUME 218

For further volumes:
http://www.springer.com/series/398

Reviews of Environmental Contamination and Toxicology

Editor
David M. Whitacre

VOLUME 218

 Springer

Coordinating Board of Editors

ISSN 0179-5953
ISBN 978-1-4899-9933-7   ISBN 978-1-4614-3137-4 (eBook)
DOI 10.1007/978-1-4614-3137-4
Springer New York Dordrecht Heidelberg London

Foreword

International concern in scientific, industrial, and governmental communities over traces of xenobiotics in foods and in both abiotic and biotic environments has justified the present triumvirate of specialized publications in this field: comprehensive reviews, rapidly published research papers and progress reports, and archival documentations. These three international publications are integrated and scheduled to provide the coherency essential for nonduplicative and current progress in a field as dynamic and complex as environmental contamination and toxicology. This series is reserved exclusively for the diversified literature on "toxic" chemicals in our food, our feeds, our homes, recreational and working surroundings, our domestic animals, our wildlife, and ourselves. Tremendous efforts worldwide have been mobilized to evaluate the nature, presence, magnitude, fate, and toxicology of the chemicals loosed upon the Earth. Among the sequelae of this broad new emphasis is an undeniable need for an articulated set of authoritative publications, where one can find the latest important world literature produced by these emerging areas of science together with documentation of pertinent ancillary legislation.

Research directors and legislative or administrative advisers do not have the time to scan the escalating number of technical publications that may contain articles important to current responsibility. Rather, these individuals need the background provided by detailed reviews and the assurance that the latest information is made available to them, all with minimal literature searching. Similarly, the scientist assigned or attracted to a new problem is required to glean all literature pertinent to the task, to publish new developments or important new experimental details quickly, to inform others of findings that might alter their own efforts, and eventually to publish all his/her supporting data and conclusions for archival purposes.

In the fields of environmental contamination and toxicology, the sum of these concerns and responsibilities is decisively addressed by the uniform, encompassing, and timely publication format of the Springer triumvirate:

Reviews of Environmental Contamination and Toxicology [Vol. 1 through 97 (1962–1986) as Residue Reviews] for detailed review articles concerned with any aspects of chemical contaminants, including pesticides, in the total environment with toxicological considerations and consequences.

Bulletin of Environmental Contamination and Toxicology (Vol. 1 in 1966) for rapid publication of short reports of significant advances and discoveries in the fields of air, soil, water, and food contamination and pollution as well as methodology and other disciplines concerned with the introduction, presence, and effects of toxicants in the total environment.

Archives of Environmental Contamination and Toxicology (Vol. 1 in 1973) for important complete articles emphasizing and describing original experimental or theoretical research work pertaining to the scientific aspects of chemical contaminants in the environment.

Manuscripts for Reviews and the Archives are in identical formats and are peer reviewed by scientists in the field for adequacy and value; manuscripts for the *Bulletin* are also reviewed, but are published by photo-offset from camera-ready copy to provide the latest results with minimum delay. The individual editors of these three publications comprise the joint Coordinating Board of Editors with referral within the board of manuscripts submitted to one publication but deemed by major emphasis or length more suitable for one of the others.

Coordinating Board of Editors

Preface

The role of *Reviews* is to publish detailed scientific review articles on all aspects of environmental contamination and associated toxicological consequences. Such articles facilitate the often complex task of accessing and interpreting cogent scientific data within the confines of one or more closely related research fields.

In the nearly 50 years since *Reviews of Environmental Contamination and Toxicology* (formerly *Residue Reviews*) was first published, the number, scope, and complexity of environmental pollution incidents have grown unabated. During this entire period, the emphasis has been on publishing articles that address the presence and toxicity of environmental contaminants. New research is published each year on a myriad of environmental pollution issues facing people worldwide. This fact, and the routine discovery and reporting of new environmental contamination cases, creates an increasingly important function for *Reviews*.

The staggering volume of scientific literature demands remedy by which data can be synthesized and made available to readers in an abridged form. *Reviews* addresses this need and provides detailed reviews worldwide to key scientists and science or policy administrators, whether employed by government, universities, or the private sector.

There is a panoply of environmental issues and concerns on which many scientists have focused their research in past years. The scope of this list is quite broad, encompassing environmental events globally that affect marine and terrestrial ecosystems; biotic and abiotic environments; impacts on plants, humans, and wildlife; and pollutants, both chemical and radioactive; as well as the ravages of environmental disease in virtually all environmental media (soil, water, air). New or enhanced safety and environmental concerns have emerged in the last decade to be added to incidents covered by the media, studied by scientists, and addressed by governmental and private institutions. Among these are events so striking that they are creating a paradigm shift. Two in particular are at the center of everincreasing-media as well as scientific attention: bioterrorism and global warming. Unfortunately, these very worrisome issues are now superimposed on the already extensive list of ongoing environmental challenges.

The ultimate role of publishing scientific research is to enhance understanding of the environment in ways that allow the public to be better informed. The term "informed public" as used by Thomas Jefferson in the age of enlightenment conveyed the thought of soundness and good judgment. In the modern sense, being "well informed" has the narrower meaning of having access to sufficient information. Because the public still gets most of its information on science and technology from TV news and reports, the role for scientists as interpreters and brokers of scientific information to the public will grow rather than diminish. Environmentalism is the newest global political force, resulting in the emergence of multinational consortia to control pollution and the evolution of the environmental ethic. Will the newpolitics of the twenty-first century involve a consortium of technologists and environmentalists, or a progressive confrontation? These matters are of genuine concern to governmental agencies and legislative bodies around the world.

For those who make the decisions about how our planet is managed, there is an ongoing need for continual surveillance and intelligent controls to avoid endangering the environment, public health, and wildlife. Ensuring safety-in-use of the many chemicals involved in our highly industrialized culture is a dynamic challenge, for the old, established materials are continually being displaced by newly developed molecules more acceptable to federal and state regulatory agencies, public health officials, and environmentalists.

Reviews publishes synoptic articles designed to treat the presence, fate, and, if possible, the safety of xenobiotics in any segment of the environment. These reviews can be either general or specific, but properly lie in the domains of analytical chemistry and its methodology, biochemistry, human and animal medicine, legislation, pharmacology, physiology, toxicology, and regulation. Certain affairs in food technology concerned specifically with pesticide and other food-additive problems may also be appropriate.

Because manuscripts are published in the order in which they are received in final form, it may seem that some important aspects have been neglected at times. However, these apparent omissions are recognized, and pertinent manuscripts are likely in preparation or planned. The field is so very large and the interests in it are so varied that the editor and the editorial board earnestly solicit authors and suggestions of underrepresented topics to make this international book series yet more useful and worthwhile.

Justification for the preparation of any review for this book series is that it deals with some aspect of the many real problems arising from the presence of foreign chemicals in our surroundings. Thus, manuscripts may encompass case studies from any country. Food additives, including pesticides, or their metabolites that may persist into human food and animal feeds are within this scope. Additionally, chemical contamination in any manner of air, water, soil, or plant or animal life is within these objectives and their purview.

Manuscripts are often contributed by invitation. However, nominations for new topics or topics in areas that are rapidly advancing are welcome. Preliminary communication with the editor is recommended before volunteered review manuscripts are submitted.

Summerfield, NC, USA   David M. Whitacre

Contents

Human Pharmaceuticals in the Aquatic Environment: A Review of Recent Toxicological Studies and Considerations for Toxicity Testing

John M. Brausch, Kristin A. Connors, Bryan W. Brooks, and Gary M. Rand

Contents

J.M. Brausch • G.M. Rand (✉)
Ecotoxicology and Risk Assessment Laboratory, Department of Earth and Environment,
Southeastern Environmental Research Center, Florida International University,
3000 NE 151st St, North Miami, FL 33181, USA
e-mail: randg@fiu.edu

K.A. Connors • B.W. Brooks
Department of Environmental Science, Center for Reservoir
and Aquatic Systems Research, Institute of Biomedical Studies, Baylor University,
Waco, TX 76798, USA

D.M. Whitacre (ed.), *Reviews of Environmental Contamination and Toxicology*,
Reviews of Environmental Contamination and Toxicology 218,
DOI 10.1007/978-1-4614-3137-4_1, © Springer Science+Business Media, LLC 2012

1 Introduction

The widespread environmental presence of human pharmaceuticals in effluents discharged from wastewater treatment plants (WWTP) has increased concern for potential ecological effects and hazard to aquatic species (Higniite and Azarnoff 1977; Halling-Sorensen et al. 1998; Daughton and Ternes 1999). Awareness of pharmaceuticals in the aquatic environment arose with the publication of two critical reviews by Halling-Sorensen et al. (1998) and Daughton and Ternes (1999), which coincided with a time of heightened concern over the presence and potential effects of endocrine active compounds in the environment (Daston et al. 1997; Vos et al. 2000) and advances in analytical detection capabilities for pharmaceuticals (Ternes et al. 2004). Furthermore, widespread detection of pharmaceuticals in the environment, ranging between ng/L and μg/L levels in surface waters and effluents and ng/kg to μg/kg levels in aquatic and terrestrial organisms, has caused increased concern and fomented new investigations of potential effects of biosolids and WWTP effluents.

Pharmaceuticals are manufactured to produce therapeutic effects in humans and animals at low concentrations. Additionally, pharmaceuticals often have physical–chemical characteristics (e.g., can pass through membranes) that are similar to many other xenobiotics. For example, they may be persistent in aquatic ecosystems (e.g., tetracycline and quinoline antibiotics) as a result of their chemical properties (Zuccato et al. 2004), or their continued replenishment from WWTP effluent (Daughton and Ternes 1999). Human pharmaceuticals have thus been classified as "pseudo-persistent" (Daughton 2002) and exhibit a longer "effective exposure duration" (Ankley et al. 2007), particularly in effluent-dominated systems (Brooks et al. 2006).

Moreover, the high polarity and low volatility of most pharmaceutical compounds increases the probability that pharmaceuticals will ultimately be transported to surface waters, if they are not removed by wastewater treatment (Breton and Boxall 2003). The potential environmental effects that pharmaceuticals pose in surface waters remain largely unknown (Jørgensen and Halling-Sørensen 2000), although it is estimated that 10–15% of pharmaceuticals found in surface waters are acutely or chronically toxic for certain endpoints, when tested in standardized aquatic toxicity models (EU [European Union] 2001; Sanderson et al. 2003).

Bioaccumulation and bioconcentration studies have been performed on approximately 30 pharmaceuticals in freshwater aquatic organisms (Daughton and Brooks 2011). Larsson et al. (1999) provided the first report of the bioconcentration in bile of a lipophilic human pharmaceutical, 17α-ethinylestradiol (EE2), while the test species, juvenile rainbow trout, were caged in a Swedish effluent-dominated river. Brooks et al. (2005) identified the potential for other human pharmaceuticals, i.e., the weak base selective serotonin reuptake inhibitors (SSRIs) fluoxetine and sertraline (more polar than EE2), to accumulate in muscle, liver, and brain tissues of three fish species collected from Pecan Creek, an effluent-dominated stream in north central Texas, United States of America (USA). Ramirez et al. (2007) advanced this work by developing a method to screen for a variety of pharmaceuticals in fish tissue, and then applied this method to identify several new classes of pharmaceuticals that could accumulate in fish from Pecan Creek.

In 2009, the U.S. Environmental Protection Agency's (EPA) National Pilot Project of Pharmaceuticals and Personal Care Products (PPCP) in Fish extended this work to five other urban river locations in the USA (Ramirez et al. 2009). Similar observations in fish tissues have been reported elsewhere (Chu and Metcalfe 2007; Kwon et al. 2009; Schultz et al. 2010). Other researchers have examined pharmaceutical concentrations in fish plasma (Brown et al. 2007; Fick et al. 2010) and have developed unique methods to examine pharmaceuticals that appear in vivo, by using solid phase microextraction technologies (Zhou et al. 2008; Zhang et al. 2010). Examining internal dose and critical tissue residues has benefit for pharmaceuticals (Brooks et al. 2009a), because internal dosimetry may be useful in estimating risks during prospective and retrospective aquatic organismal risk assessments (Huggett et al. 2003; Fick et al. 2010; Berninger and Brooks 2010; Berninger et al. 2011). Unfortunately, information on accumulation of pharmaceuticals in estuarine and marine organisms is decidedly lacking (Daughton and Brooks 2011). Although a better understanding of the pharmacokinetics of pharmaceuticals in aquatic organisms is needed, studies to date have emphasized the importance of understanding the relationships that exist among the magnitude, frequency and duration of field exposures, internal dosimetry and adverse effects thresholds.

Although the number of publications that concern the potential toxicity of pharmaceuticals to aquatic organisms have increased (Crane et al. 2006; Fent et al. 2006; Brooks et al. 2009a; Santos et al. 2010), there is still a dearth of information on chronic exposures and exposures to pharmaceutical mixtures in effluent (Cleuvers 2003, 2004). Santos et al. (2010) compiled data from 94 articles published between 1996 and 2009, which disclosed that almost 70% of published reports dealt with acute toxicity. Ecotoxicological data, published either in the peer-reviewed literature or in ecotoxicological databases (ECE TOX in EU; ECOTOX in USA) (EU 2001; Jones et al. 2002; Sanderson et al. 2003; Stuer-Lauridsen et al. 2000), are currently available for <10% of the currently prescribed pharmaceuticals. It is also disconcerting that only a few pharmaceuticals have been subjected to ecological risk assessments (Henschel et al. 1997; Emblidge and DeLorenzo 2006; Hernando et al. 2006; Küster et al. 2010; Oakes et al. 2010; Liebig et al. 2010). Furthermore, the majority of aquatic toxicity studies performed on pharmaceuticals address the effects of individual-substance exposures in laboratory waters (or a few with sediments), rather than effluent matrices that reflect how pharmaceuticals are typically discharged to surface waters. Of the limited number of microcosm/mesocosm studies that have been performed on pharmaceuticals, none have focused on experimental streams, which are the primary receiving systems for pharmaceutical discharges.

Over the last 15 years, regulatory agencies have started to develop guidelines for analyzing pharmaceuticals in environmental compartments and for testing their potential toxic effects in the environment (European Medicines Agency [EMEA] 1998; Food and Drug Administration-Center for Drug Evaluation and Research [FDA-CDER] 1998; International Cooperation on Harmonisation of Technical Requirements for Registration of Veterinary Products [VICH] 2000, 2004). The EU released guidelines in 1998 for analyzing veterinary pharmaceuticals (EMEA 1998), and supplemented these in 2005 by incorporating human-use pharmaceuticals (EMEA 2005). For veterinary medicines, phase I and II testing guidelines were expanded in 2000 (VICH 2000) and 2004 (VICH 2004), respectively. Brooks et al. (2008, 2009b) further examined aquatic toxicology and risk assessment approaches for veterinary medicines, using VICH guidelines and other criteria. In the USA, the USFDA also published guidelines in 1998 on the analysis of pharmaceutical active ingredients, but these were only cogent for compounds having expected environmental concentrations >1 μg/L (FDA-CDER 1998).

The objective of this review of the peer-reviewed literature is to summarize and provide an overview of the current status of freshwater aquatic toxicity testing of human pharmaceuticals. Moreover, we have endeavored to identify data gaps in the literature and provide our recommendations on what aquatic toxicity tests are appropriate for evaluating the safety of pharmaceuticals in wastewaters. The scope of the literature reviewed herein, includes the representative standardized and nontraditional acute and chronic aquatic toxicity studies that were performed on aquatic organisms and were published through January 2011. It is our intent and our hope that the recommendations provided herein for aquatic toxicity testing will support future assessments of pharmaceuticals in freshwater systems.

2 Acute Toxicity of Pharmaceuticals in Aquatic Organisms

The acute toxicity of pharmaceuticals has been studied for over 150 individual compounds that collectively comprise 35 pharmaceutical classes. Data for each of these compounds are presented in Table 1. The organisms from the majority of the main

Table 1 Acute toxicity data for human-use pharmaceuticals[a]

Compound	Category	Species	Trophic group	Endpoint/Duration[a]	Value (mg/L)	References
Acetylsalicylic acid (aspirin)	Analgesic	H. vulgaris	Invert.	7-day EC$_{50}$	>100	Pascoe et al. (2003)
		D. magna	Invert.	24-h EC$_{50}$	1,468	Lilius et al. (1994)
		D. magna	Invert.	24-h EC$_{50}$	168	Calleja et al. (1994)
		D. magna	Invert.	48-h EC$_{50}$	88.1	Cleuvers (2004)
		S. proboscideus	Invert.	24-h LC$_{50}$	178	Calleja et al. (1994)
		B. calyciflorus	Invert.	24-h LC$_{50}$	141	Calleja et al. (1994)
		D. subspicatus	Algae	96-h EC$_{50}$	106.7	Cleuvers (2004)
Budesonide	Analgesic	Daphnia spp.	Invert.	EC$_{50}$	20	FDA-CDER (1996)
		Unspecified fish	Fish	LC$_{50}$	>19	FDA-CDER (1996)
Dextropropoxyphene	Analgesic	D. magna	Invert.	24-h EC$_{50}$	14.6	Lilius et al. (1994)
		D. magna	Invert.	24-h EC$_{50}$	19	Calleja et al. (1994)
		S. proboscideus	Invert.	24-h LC$_{50}$	7.6	Calleja et al. (1994)
		B. calyciflorus	Invert.	24-h LC$_{50}$	42	Calleja et al. (1994)
Diclofenac	Analgesic	V. fischeri	Bacteria	EC$_{50}$	11.45	Ferrari et al. (2003)
		C. dubia	Invert.	48-h EC$_{50}$	22.7	Ferrari et al. (2003)
		D. magna	Invert.	48-h EC$_{50}$	68	Cleuvers (2003)
		D. magna	Invert.	48-h EC$_{50}$	22.4	Ferrari et al. (2003)
		T. battagliai	Invert.	48-h LC$_{50}$	15.8	Schmidt et al. (2011)
		S. costatum	Algae	72-h IC$_{50}$	5	Schmidt et al. (2011)
		D. subspicatus	Algae	3-day EC$_{50}$	72	Cleuvers (2003)
		P. subsapitata	Algae	96-h LOEC	20	Ferrari et al. (2003)
		S. leopoliensis	Algae	96-h LOEC	10	Ferrari et al. (2003)
		C. meneghiniana	Algae	96-h LOEC	10	Ferrari et al. (2003)
		D. tertolecta	Algae	96-h EC$_{50}$	185.7	Lin et al. (2009)
		D. rerio(embryo)	Fish	72-h LC$_{50}$	7.8	van den Brandof and Montforts (2010)

(continued)

Table 1 (continued)

Compound	Category	Species	Trophic group	Endpoint/Duration[a]	Value (mg/L)	References
Ibuprofen	Analgesic	Biofilm		EC$_{50}$	>0.01	Lawrence et al. (2005)
		V. fischeri	Bacteria	EC$_{50}$	12.3	Halling-Sorensen et al. (1998)
		D. magna	Invert.	48-h EC$_{50}$	108	Cleuvers (2003)
		D. magna	Invert.	48-h EC$_{50}$	9.06	Knoll/BASF (1995)
		H. vulgaris	Invert.	EC$_{50}$	>100	Pascoe et al. (2003)
		T. platyurus	Invert.	24-h LC$_{50}$	19.59	Kim et al. (2009)
		H. attenuata	Invert.	96-h LC$_{50}$, EC$_{50}$	22.36, 1.65 (Morph.)	Quinn et al. (2008)
		C. vaiegatus	Fish	NEL	>300	Knoll/BASF (1995)
		L. macrochirus	Fish	96-h LC$_{50}$	173	Knoll/BASF (1995)
		O. latipes	Fish	96-h LC$_{50}$	>100	Pounds et al. (2008)
		P. carinatus	Mollusc	72-h LC$_{50}$	17.1	Pounds et al. (2008)
		X. laevis	Amphib.	EC$_{10}$	30.7	Richards and Cole (2006)
		S. capricornutum	Algae	96-h NEL	>30	Knoll/BASF (1995)
		D. subspicatus	Algae	72-h EC$_{50}$	315	Lawrence et al. (2005)
		Synechocystis sp.	Algae	120-h LOEC	1	Pomati et al. (2004)
		S. costatum	Algae	96-h EC$_{50}$	7.1	Knoll/BASF (1995)
Indomethacin	Analgesic	T. platyurus	Invert.	24-h LC$_{50}$	16.14	Kim et al. (2009)
		O. latipes	Fish	96-h LC$_{50}$	81.92	Kim et al. (2009)
Ketorolac	Analgesic	L. macrochirus	Fish	96-h LC$_{50}$	1480	Anon 1993
Mefenamic acid	Analgesic	T. platyurus	Invert.	24-h LC$_{50}$	3.95	Kim et al. (2009)
		O. latipes	Fish	96-h LC$_{50}$	8.04	Kim et al. (2009)
Naproxen sodium	Analgesic	D. magna	Invert.	24-h EC$_{50}$	140	Rodriguez et al. (1992)
		D. magna	Invert.	48-h EC$_{50}$	174	Cleuvers (2003)
		D. magna	Invert.	48-h EC$_{50}$	166.3	Cleuvers (2004)
		C. dubia	Invert.	24-h EC$_{50}$	66.37	Isidori et al. (2005a)
		B. calyciflorus	Invert.	24-h LC$_{50}$	62.48	Isidori et al. (2005a)
		T. platyurus	Invert.	24-h LC$_{50}$	84.09	Isidori et al. (2005a)
		H. azteca	Invert.	96-h LC$_{50}$	383	Rodriguez et al. (1992)
		L. macrochirus	Fish	96-h LC$_{50}$	560	Rodriguez et al. (1992)

Compound	Class	Species	Organism	Endpoint	Value	Reference
Paracetamol/ acetominophen	Analgesic	Oncorhynchus mykiss	Fish	96-h LC$_{50}$	690	Rodriguez et al. (1992)
		D. subspicatus	Algae	72-h EC$_{50}$	>320	Cleuvers (2003)
		P. subcapitata	Algae	72-h EC$_{50}$	31.8	Isidori et al. (2005b)
		D. subspicatus	Algae	96-h EC$_{50}$	625.5	Cleuvers (2004)
		V. fischeri	Bacteria	LC$_{50}$	548.7	Kim et al. (2007)
		V. fischeri	Bacteria	EC$_{50}$	567.5	Kim et al. (2007)
		D. magna	Invert.	48-h LC$_{50}$	30.1	Kim et al. (2007)
		D. magna	Invert.	96-h EC$_{50}$	26.6	Kim et al. (2007)
		D. magna	Invert.	24-h EC$_{50}$	555	Calleja et al. (1994)
		D. magna	Invert.	24-h EC$_{50}$	13	Kuhn et al. 1989
		D. magna	Invert.	48-h EC$_{50}$	92	Kuhn et al. 1989
		D. magna	Invert.	24 EC$_{50}$	293	Henschel et al. (1997)
		D. magna	Invert.	48 EC$_{50}$	50.0	Henschel et al. (1997)
		S. proboscideus	Invert.	24-h LC$_{50}$	29.6	Calleja et al. (1994)
		B. calyciflorus	Invert.	24-h LC$_{50}$	5,306	Calleja et al. (1994)
		T. pyriformis	Invert.	48-h EC$_{50}$	112	Henschel et al. (1997)
		D. rerio (embryos)	Fish	48-h EC$_{50}$	378	Henschel et al. (1997)
		O. latipes	Fish	48-h LC$_{50}$	>160	Kim et al. (2007)
		X. laevis	Amphib.	96-h EC$_{10}$	>100	Richards and Cole (2006)
		S. subspicatus	Algae	72-h EC$_{50}$	134	Henschel et al. (1997)
Salicylic acid	Analgesic	T. pyriformis	Protozoa	48-h	>100	Bringmann and Kuhn (1982)
		D. magna	Invert.	24-h EC$_{50}$	>1440	Bringmann and Kuhn (1982)
		D. magna	Invert.	24-h EC$_{50}$	230	Wang and Lay (1989)
		D. magna	Invert.	EC$_{50}$	118	Henschel et al. (1997)
		B. rerio (embryos)	Fish	48-h EC$_{50}$	37.0	Henschel et al. (1997)
		S. subspicatus	Algae	72-h EC$_{50}$	>100	Henschel et al. (1997)
Tramadol	Analgesic	Unspecified fish	Fish	LC$_{50}$	130	FDA-CDER (1996)
		Daphnia spp	Invert.	EC$_{50}$	73	FDA-CDER (1996)
Tolazoline	Antiadrenergic	P. promelas	Fish	96-h EC$_{50}$	354	Russom et al. (1997)

(continued)

Table 1 (continued)

Compound	Category	Species	Trophic group	Endpoint/Duration[a]	Value (mg/L)	References
Bicalutamide	Anti-androgen	Daphnia spp.	Invert.	EC_{50}	>5	FDA-CDER (1996)
		Unspecified green algae	Algae	EC_{50}	>1	FDA-CDER (1996)
		Unspecified blue-green algae	Algae	EC_{50}	>1	FDA-CDER (1996)
Finasteride	Anti-androgen	Daphnia spp.	Invert.	EC_{50}	21	FDA-CDER (1996)
		O. mykiss	Fish	LC_{50}	20	FDA-CDER (1996)
Flutamide	Anti-androgen	P. promelas	Fish	14-day LC_{50}	>1,000	Panter et al. (1999)
Quinidine sulfate	Anti-arrhythmic	D. magna	Invert.	24-h EC_{50}	60	Lilius et al. (1994)
		D. magna	Invert.	24-h EC_{50}	60	Calleja et al. (1994)
		S. proboscideus	Invert.	24-h LC_{50}	8.3	Calleja et al. (1994)
		B. calyciflorus	Invert.	24-h LC_{50}	8.7	Calleja et al. (1994)
Verapamil	Anti-arrhythmic	D. magna	Invert.	24-h EC_{50}	327	Lilius et al. (1994)
		D. magna	Invert.	24-h EC_{50}	55.5	Calleja et al. (1994)
		S. proboscideus	Invert.	24-h LC_{50}	6.24	Calleja et al. (1994)
		B. calyciflorus	Invert.	24-h LC_{50}	10.90	Calleja et al. (1994)
		O. mykiss (juvenile)	Fish	96-h LC_{50}	2.72	Li et al. (2010)
Fluticasone	Anti-asthmatic	Daphnia spp.	Invert.	EC_{50}	0.55	FDA-CDER (1996)
Salmeterol	Anti-asthmatic	Daphnia spp.	Invert.	EC_{50}	20	
Theophylline	Anti-asthmatic	D. magna	Invert.	24-h EC_{50}	155	Lilius et al. (1994)
		D. magna	Invert.	24-h EC_{50}	483	Calleja et al. (1994)
		S. proboscideus	Invert.	24-h LC_{50}	425	Calleja et al. (1994)
		B. calyciflorus	Invert.	24-h LC_{50}	3,926	Calleja et al. (1994)
Acriflavine	Antibiotic (Antiseptic)	M. saxatilis (larvae)	Fish	96-h LC_{50}	5	Hughes (1973)
		M. saxatilis (fingerling)	Fish	96-h LC_{50}	27.5	Hughes (1973)
		O. mykiss	Fish	24-h LC_{50}	30.1	Wilford (1966)
		O. mykiss	Fish	48-h LC_{50}	19.9	Wilford (1966)

		Species	Group	Endpoint	Value	Reference
		S. namaycush	Fish	24-h LC$_{50}$	37.5	Wilford (1966)
		S. namaycush	Fish	48-h LC$_{50}$	28.0	Wilford (1966)
		S. trutta	Fish	24-h LC$_{50}$	40.0	Wilford (1966)
		S. trutta	Fish	48-h LC$_{50}$	27.0	Wilford (1966)
		I. punctatus	Fish	24-h LC$_{50}$	43.5	Wilford (1966)
		I. punctatus	Fish	48-h LC$_{50}$	33.2	Wilford (1966)
		S. fontinalls	Fish	24-h LC$_{50}$	48.0	Wilford (1966)
		S. fontinalls	Fish	48-h LC$_{50}$	14.8	Wilford (1966)
		L. macrochirus	Fish	24-h LC$_{50}$	18.0	Wilford (1966)
		L. macrochirus	Fish	48-h LC$_{50}$	13.5	Wilford (1966)
Aminosidine sulfate (neomycin E)	Antibiotic (antiamebic)	D. magna	Invert.	24-h LC$_{50}$	1,055	Di Delupis et al. (1992)
		D. magna	Invert.	48-h LC$_{50}$	503	Di Delupis et al. (1992)
Amopyroquin dihydrochloride	Antibiotic (antimalarial)	O. mykiss	Fish	24-h LC$_{50}$	47.0	Wilford (1966)
		O. mykiss	Fish	48-h LC$_{50}$	35.3	Wilford (1966)
		S. namaycush	Fish	24-h LC$_{50}$	15.5	Wilford (1966)
		S. namaycush	Fish	48-h LC$_{50}$	140	Wilford (1966)
		S. trutta	Fish	24-h LC$_{50}$	42.0	Wilford (1966)
		S. trutta	Fish	48-h LC$_{50}$	360	Wilford (1966)
		I. punctatus	Fish	24-h LC$_{50}$	19.8	Wilford (1966)
		I. punctatus	Fish	48-h LC$_{50}$	12.5	Wilford (1966)
		S. fontinalis	Fish	24-h LC$_{50}$	52.0	Wilford (1966)
		S. fontinalis	Fish	48-h LC$_{50}$	40.0	Wilford (1966)
		L. macrochirus	Fish	24-h LC$_{50}$	330	Wilford (1966)
		L. macrochirus	Fish	48-h LC$_{50}$	18.5	Wilford (1966)
Amoxicillin	Antibiotic	V. fischeri	Bacteria	15-min EC$_{50}$	3.597	Park and Cjoi (2008)
		H. vulgaris	Invert.	EC$_{50}$	>100	Pascoe et al. (2003)
		M. aeruginosa	Algae	72-h EC$_{50}$	0.0037	Hulton et al. (1999)
		S. capricornutum	Algae	72-h NOEC	250,000	Hulton et al. (1999)
		S. leopoliensis	Algae	72-h EC$_{50}$; LOEC, NOEC	0.00222, 0.00156, 0.00078	Andreozzi et al. (2004)

(continued)

Table 1 (continued)

Compound	Category	Species	Trophic group	Endpoint/Duration[a]	Value (mg/L)	References
Ampicillin	Antibiotic	V. fischeri	Bacteria	15-min EC_{50}	2,627	Park and Cjoi (2008)
Azithromycin	Antibiotic	Unspecified amphipod	Invert.	LC_{50}	>120	FDA-CDER (1996)
		Daphnia spp.	Invert.	EC_{50}	120	FDA-CDER (1996)
Bacitracin	Antibiotic	D. magna	Invert.	24-h LC_{50}	126.4	Di Delupis et al. (1992)
		D. magna	Invert.	24-h LC_{50}	30.5	Di Delupis et al. (1992)
		D. magna	Invert.	24-h LC_{50}	126.4	Brambilla et al. (1994)
		D. magna	Invert.	48-h LC_{50}	30.5	Brambilla et al. (1994)
Cefprozil	Antibiotic	Daphnia spp.	Invert.	EC_{50}	>642	FDA-CDER (1996)
Ceftibuten	Antibiotic	Daphnia spp.	Invert.	EC_{50}	>600	FDA-CDER (1996)
		Amphipod	Invert.	LC_{50}	>520	FDA-CDER (1996)
Chloramine T	Antibiotic	R. heteromorpha	Fish	96-h LC_{50}	22	Tooby et al. (1975)
Chloramphenicol	Antibiotic	D. magna	Invert.	24-h EC_{50}	543	Lilius et al. (1994)
		D. magna	Invert.	24-h EC_{50}	1,086	Calleja et al. (1994)
		S. proboscideus	Invert.	24-h LC_{50}	305	Calleja et al. (1994)
		B. calyciflorus	Invert.	24-h LC_{50}	2074	Calleja et al. (1994)
Chloroquine	Antibiotic (Antimalarial)	D. magna	Invert.	24-h EC_{50}	50	Lilius et al. (1994)
		D. magna	Invert.	24-h EC_{50}	43.5	Calleja et al. (1994)
		S. proboscideus	Invert.	24-h LC_{50}	11.7	Calleja et al. (1994)
		B. calyciflorus	Invert.	24-h LC_{50}	4.39	Calleja et al. (1994)
Clarithromycin	Antibiotic	D. magna	Invert.	24-h EC_{50}	25.72	Isidori et al. (2005b)
		C. dubia	Invert.	24-h EC_{50}	18.66	Isidori et al. (2005b)
		T. platyurus	Invert.	24-h LC_{50}	33.64	Isidori et al. (2005b)
		T. platyurus	Invert.	24-h LC_{50}	94.23	Kim et al. (2009)
		B. calyciflorus	Invert.	24-h LC_{50}	35.46	Isidori et al. (2005b)
		O. latipes	Fish	96-h LC_{50}	>100	Kim et al. (2009)
		P. subcapitata	Algae	96-h EC_{50}, LOEC	0.011, 0.0063, 0.0031	Yamashita et al. (2006)
		P. subcapitata	Algae	72-h EC_{50}	0.002	Isidori et al. (2005b)

Compound	Class	Species	Group	Endpoint	Value	Reference
Chlortetracycline	Antibiotic	V. fischeri	Bacteria	15-min EC$_{50}$	13	Park and Cjoi (2008)
		D. magna	Invert.	24, 48-h EC$_{50}$	380.1, 225	Park and Cjoi (2008)
		M. macrocopa	Invert.	24, 48-h EC$_{50}$	515, 272	Park and Cjoi (2008)
		O. latipes	Fish	24, 48-h LC$_{50}$	88.4, 78.9	Park and Cjoi (2008)
		M. aeruginosa	Algae	EC$_{50}$	0.05	Halling-Sorensen (2000)
		S. capricornutum	Algae	EC$_{50}$	3.1	Halling-Sorensen (2000)
Ciprofloxacin	Antibiotic	D. magna	Invert.	48-h LC$_{50}$	>100	Robinson et al. (2005)
		P. promelas	Fish	7-day EC$_{50}$	10	Robinson et al. (2005)
		X. laevis	Amphib.	96-h EC$_{50}$	>100	Richards and Cole (2006)
		M. aeruginosa	Algae	120-h EC$_{50}$	17	Robinson et al. (2005)
		P. subcapitata	Algae	72-h EC$_{50}$	18.7	Robinson et al. (2005)
Clinafloxacin	Antibiotic	D. magna	Invert.	48-h LC$_{50}$	>10	Robinson et al. (2005)
		P. promelas	Fish	7-day EC$_{50}$	>10	Robinson et al. (2005)
		M. aeruginosa	Algae	120-h EC$_{50}$	0.1	Robinson et al. (2005)
		P. subcapitata	Algae	72-h EC$_{50}$	1.1	Robinson et al. (2005)
Cyclosporine	Antibiotic	Daphnia spp.	Invert.	EC$_{50}$	20	FDA-CDER (1996)
		O. mykiss	Fish	LC$_{50}$	>100	FDA-CDER (1996)
Didanosine	Antibiotic (retroviral)	D. magna	Invert.	EC$_{50}$	>1,020	FDA-CDER (1996)
Dirithromycin	Antibiotic	D. magna	Invert.	EC$_{50}$	>48	FDA-CDER (1996)
		O. mykiss	Fish	LC$_{50}$	>2,880	FDA-CDER (1996)
Enrofloxacin	Antibiotic	V. fischeri	Bacteria	15-min EC$_{50}$	326.8	Park and Cjoi (2008)
		D. magna	Invert.	48-h LC$_{50}$	>10	Robinson et al. (2005)
		D. magna	Invert.	24, 48-h EC$_{50}$	131.7, 56.7	Park and Cjoi (2008)
		M. macrocopa	Invert.	24-h EC$_{50}$	285.7	Park and Cjoi (2008)
		M. aeruginosa	Algae	120-h EC$_{50}$	0.05	Robinson et al. (2005)
		P. subcapitata	Algae	72-h EC$_{50}$	3.1	Robinson et al. (2005)
		P. promelas	Fish	7-day EC$_{50}$	>10	Robinson et al. (2005)

(continued)

Table 1 (continued)

Compound	Category	Species	Trophic group	Endpoint/Duration[a]	Value (mg/L)	References
Erythromycin	Antibiotic	D. magna	Invert.	24-h LC_{50}	388	Di Delupis et al. (1992)
		D. magna	Invert.	48-h LC_{50}	211	Di Delupis et al. (1992)
		D. magna	Invert.	24-h EC_{50}	22.45	Isidori et al. (2005b)
		C. dubia	Invert.	24-h EC_{50}	10.23	Isidori et al. (2005b)
		T. platyurus	Invert.	24-h LC_{50}	>100	Kim et al. (2009)
		T. platyurus	Invert.	24-h LC_{50}	17.68	Isidori et al. (2005b)
		B. calyciflorus	Invert.	24-h LC_{50}	27.53	Isidori et al. (2005b)
		O. latipes	Fish	96-h LC_{50}	>100	Kim et al. (2009)
		P. subcapitata	Algae	72-h EC_{50}	0.020	Isidori et al. (2005b)
		S. capricornutum	Algae	72-h EC_{50}, NOEC	0.0366, 0.0103	Costanzo et al. (2005)
		C. vulgaris	Algae	72-h EC_{50}, NOEC	33.8, 12.5	Costanzo et al. (2005)
Erythromycin phosphate	Antibiotic	S. namaycush	Fish	24-h LC_{50}	818	Marking et al. (1988)
		S. namaycush	Fish	96-h LC_{50}	410	Marking et al. (1988)
Erythromycin thiocyanate	Antibiotic	O. mykiss	Fish	48-h LC_{50}	>80	Wilford (1966)
		S. trutta	Fish	48-h LC_{50}	>80	Wilford (1966)
		S. fontinalis	Fish	48-h LC_{50}	>80	Wilford (1966)
		I. punctatus	Fish	48-h LC_{50}	>80	Wilford (1966)
		L. macrochirus	Fish	48-h LC_{50}	>80	Wilford (1966)
		S. namaycush	Fish	48-h LC_{50}	>80	Wilford (1966)
		Synechocystis sp.	Algae	120-h LOEC	1	Pomati et al. (2004)
Famciclovir	Antibiotic (Retroviral)	D. magna	Invert.	EC_{50}	820	FDA-CDER (1996)
		L. macrochirus	Fish	LC_{50}	>986	FDA-CDER (1996)
Flumequine	Antibiotic	D. magna	Invert.	48-h LC_{50}	>10	Robinson et al. (2005)
		P. promelas	Fish	7-day EC_{50}	>10	Robinson et al. (2005)
		M. aeruginosa	Algae	120-h EC_{50}	1.96	Robinson et al. (2005)
		P. subcapitata	Algae	72-h EC_{50}	5	Robinson et al. (2005)
Isoniazid	Antibiotic	D. magna	Invert.	24-h EC_{50}	85	Lilius et al. (1994)
		D. magna	Invert.	24-h EC_{50}	125.5	Calleja et al. (1994)

Drug	Type	Species	Endpoint	Value	Reference
Levofloxacin	Antibiotic	S. proboscideus	Invert.	24-h LC$_{50}$	24.4	Calleja et al. (1994)
		B. calyciflorus	Invert.	24-h LC$_{50}$	3.045	Calleja et al. (1994)
		D. magna	Invert.	48-h LC$_{50}$	>10	Robinson et al. (2005)
		T. platyurus	Invert.	24-h LC$_{50}$	>100	Kim et al. (2009)
		P. promelas	Fish	7-day LC$_{50}$	10	Robinson et al. (2005)
		O. latipes	Fish	96-h LC$_{50}$	>100	Kim et al. (2009)
		X. laevis	Amphib.	96-h EC$_{50}$	>100	Richards and Cole (2006)
		M. aeruginosa	Algae	120-h EC$_{50}$	0.008	Robinson et al. (2005)
		P. subcapitata	Algae	72-h EC$_{50}$	7.4	Robinson et al. (2005)
Lincomycin	Antibiotic	D. magna	Invert.	72-h LC$_{50}$	379.39	Di Delupis et al. (1992)
		D. magna	Invert.	24-h EC$_{50}$	23.18	Isidori et al. (2005b)
		C. dubia	Invert.	24-h EC$_{50}$	13.98	Isidori et al. (2005b)
		T. platyurus	Invert.	24-h LC$_{50}$	30.00	Isidori et al. (2005b)
		B. calyciflorus	Invert.	24-h LC$_{50}$	24.94	Isidori et al. (2005b)
		P. subcapitata	Algae	72-h EC$_{50}$	70	Dorne et al. (2007)
Lomefloxacin	Antibiotic	D. magna	Invert.	48-h LC$_{50}$	>10	Robinson et al. (2005)
		Daphnia spp.	Invert.	EC$_{50}$	130	FDA-CDER (1996)
		P. promelas	Fish	7-day LC$_{50}$	10	Robinson et al. (2005)
		O. mykiss	Fish	LC$_{50}$	170	FDA-CDER (1996)
		M. aeruginosa	Algae	120-h EC$_{50}$	0.19	Robinson et al. (2005)
		P. subcapitata	Algae	72-h EC$_{50}$	22.7	Robinson et al. (2005)
		Unspecified green algae	Algae	EC$_{50}$	2.4	FDA-CDER (1996)
Loracarbef	Antibiotic (Antiseptic)	Daphnia spp.	Invert.	EC$_{50}$	>963	FDA-CDER (1996)
Loracarbef	Antibiotic (Antiseptic)	O. mykiss	Fish	24-h LC$_{50}$	60.5	Wilford (1966)
		O. mykiss	Fish	48-h LC$_{50}$	21.2	Wilford (1966)
		S. namaycush	Fish	24-h LC$_{50}$	13.0	Wilford (1966)
		S. namaycush	Fish	48-h LC$_{50}$	2.13	Wilford (1966)

(continued)

Table 1 (continued)

Compound	Category	Species	Trophic group	Endpoint/Duration[a]	Value (mg/L)	References
		S. trutta	Fish	24-h LC$_{50}$	110	Wilford (1966)
		S. trutta	Fish	48-h LC$_{50}$	54.0	Wilford (1966)
		I. punctatus	Fish	24-h LC$_{50}$	750	Wilford (1966)
		I. punctatus	Fish	48-h LC$_{50}$	5.65	Wilford (1966)
		S. fontinalis	Fish	24-h LC$_{50}$	895	Wilford (1966)
		S. fontinalis	Fish	48-h LC$_{50}$	745	Wilford (1966)
		L. macrochirus	Fish	24-h LC$_{50}$	110	Wilford (1966)
		L. macrochirus	Fish	48-h LC$_{50}$	645	Wilford (1966)
Metronidazole	Antibiotic (antiprotozoal)	D. magna	Invert.	48-h LOEC	1,000	Wollenberger et al. (2000)
		B. rerio	Fish	96-h EC$_{50}$	>500	Lanzky and Halling-Sorenson (1997)
		O. mykiss	Fish	48-h LC$_{50}$	>100	Wilford (1966)
		S. trutta	Fish	48-h LC$_{50}$	>100	Wilford (1966)
		S.fontinalis	Fish	48-h LC$_{50}$	>100	Wilford (1966)
		I. punctatus	Fish	48-h LC$_{50}$	>100	Wilford (1966)
		L. macrochirus	Fish	48-h LC$_{50}$	>100	Wilford (1966)
		S. namaycush	Fish	48-h LC$_{50}$	>100	Wilford (1966)
		S. capricornutum	Algae	72-h EC$_{50}$	39.1	Lanzky and Halling-Sorenson (1997)
		Chiarello spp.	Algae	72-h EC$_{50}$	12.5	Lanzky and Halling-Sorenson (1997)
Nitrofurazone	Antibiotic	D. magna	Invert.	LC$_{50}$	28.7	Macri and Sbardella (1984)
		M. saxatilis (larvae)	Fish	96-h LC$_{50}$	10	Hughes (1973)
		S. capricornutum	Algae	EC$_{50}$	1.45	Macri and Sbardella (1984)
Norfloxacin	Antibiotic	B. calyciflorus	Invert.	24-h LC$_{50}$	29.88	Isidori et al. (2005b)
		S. obliquus	Algae	48-h EC$_{50}$	38.5	Nie et al. (2009)
		S. capricornutum	Algae	96-h EC$_{50}$	16.6	Eguchi et al. (2004)
		C. vulgaris	Algae	96-h EC$_{50}$	10.4	Eguchi et al. (2004)

Drug	Class	Species	Organism	Test	Value	Reference
Ofloxacin	Antibiotic	D. magna	Invert.	48-h LC_{50}	>10	Robinson et al. (2005)
		D. magna	Invert.	24-h EC_{50}	31.75	Isidori et al. (2005b)
		C. dubia	Invert.	24-h EC_{50}	17.41	Isidori et al. (2005b)
		T. platyurus	Invert.	24-h LC_{50}	33.98	Isidori et al. (2005b)
		P. promelas	Fish	7-day EC_{50}	10	Robinson et al. (2005)
		Chlorella sp.	Algae	72-h LC_{50}	2.0	Lanzky and Halling-Sorenson (1997)
		S. capricornutum	Algae	72-h LC_{50}	40.4	Lanzky and Halling-Sorenson (1997)
		P. subcapitata	Algae	72-h EC_{50}	1.44	Isidori et al. (2005b)
		P. subcapitata	Algae	72-h EC_{50}	12.1	Robinson et al. (2005)
		P. subcapitata	Algae	96-h EC_{50}	2.5	Ferrari et al. (2003)
		S. leopolensis	Algae	96-h EC_{50}	0.05	Ferrari et al. (2003)
		C. meneghiniana	Algae	96-h EC_{50}	0.31	Ferrari et al. (2003)
		M. aeruginosa	Algae	120-h growth	0.21	Robinson et al. (2005)
Oxolinic acid	Antibiotic	D. magna	Invert.	24, 48-h EC_{50}	5.9, 4.6	Wollenberger et al. (2000)
		M. aeruginosa	Algae	72-h EC_{50}	0.18	Hulton et al. (1999)
		S. capricornutum	Algae	72-h EC_{50}	16	Hulton et al. (1999)
Oxytetracycline	Antibiotic	V. fischeri	Bacteria	15-min EC_{50}	87	Park and Cjoi (2008)
		D. magna	Invert.	24-h EC_{50}	22.64	Isidori et al. (2005b)
		D. magna	Invert.	48-h LOEC	100	Wollenberger et al. (2000)
		D. magna	Invert.	24, 48-h EC_{50}	831.6, 621.2	Park and Cjoi (2008)
		C. dubia	Invert.	24-h EC_{50}	18.65	Isidori et al. (2005b)
		T. platyurus	Invert.	24-h EC_{50}	25.00	Isidori et al. (2005b)
		M. macrocopa	Invert.	24, 48-h EC_{50}	137.1, 126.7	Park and Cjoi (2008)
		B. calyciflorus	Invert.	24-h LC_{50}	34.21	Isidori et al. (2005b)
		H. attenuata	Invert.	96-h LC_{50}, EC_{50}, LOEC, NOEC	>100, 40.13, 100, 50	Quinn et al. (2008)
		O. latipes	Fish	24, 48-h LC_{50}	215.4, 110.1	Park and Cjoi (2008)
		M. saxatilis (larvae)	Fish	24-h LC_{50}	62.5	Hughes (1973)
		M. saxatilis (fingerling)	Fish	24-h LC_{50}	150	Hughes (1973)

(continued)

Table 1 (continued)

Compound	Category	Species	Trophic group	Endpoint/Duration[a]	Value (mg/L)	References
		M. saxatilis (fingerling)	Fish	48-h LC_{50}	125	Hughes (1973)
		M. saxatilis (fingerling)	Fish	72-h LC_{50}	100	Hughes (1973)
		M. saxatilis (fingerling)	Fish	96-h LC_{50}	75	Hughes (1973)
		S. namaycush	Fish	24/96-h LC_{50}	<200	Marking et al. (1988)
		M. aeruginosa	Algae	EC_{50}	0.231	Holten Lutzhoft et al. (1999)
		M. aeruginosa	Algae	72-h EC_{50}	207	Hulton et al. (1999)
		S. capricornutum	Algae	72-h EC_{50}	4,500	Hulton et al. (1999)
		S. capricornutum	Algae	EC_{50}, NOEC	342, 183	Costanzo et al. (2005)
		S. capricornutum	Algae	EC_{50}	50	Holten Lutzhoft et al. (1999)
		P. subcapitata	Algae	EC_{50}	170	Isidori et al. (2005b)
		C. vulgaris	Algae	72-h EC_{50}	7,050, <3,580	Costanzo et al. (2005)
		Rhadomonas spp.	Algae	EC_{50}, NOEC	1.7	Holten Lutzhoft et al. (1999)
		L. minor	Plant	48-h EC_{50}	4.92	Pro et al. (2003)
Penicillin	Antibiotic	M. aeruginosa	Algae	96-h EC_{50}	0.006	Halling-Sorensen (2000)
		S. capricornutum	Algae	96-h NOEC	100	Halling-Sorensen (2000)
Quinacrine	Antibiotic	O. mykiss	Fish	48-h LC_{50}	122	Wilford (1966)
		S. namaycush	Fish	24-h LC_{50}	25.0	Wilford (1966)
		S. namaycush	Fish	48-h LC_{50}	21.0	Wilford (1966)
		S. trutta	Fish	24-h LC_{50}	300	Wilford (1966)
		S. trutta	Fish	48-h LC_{50}	230	Wilford (1966)
		I. punctatus	Fish	24-h LC_{50}	196	Wilford (1966)
		I. punctatus	Fish	48-h LC_{50}	70	Wilford (1966)
		S. fontinalis	Fish	48-h LC_{50}	230	Wilford (1966)
		L. macrochirus	Fish	24-h LC_{50}	120	Wilford (1966)
		L. macrochirus	Fish	48-h LC_{50}	79	Wilford (1966)
Quinine	Antibiotic	O. mykiss	Fish	48-h LC_{50}	>100	Wilford (1966)
		S. trutta	Fish	48-h LC_{50}	>100	Wilford (1966)

Compound	Class	Species	Group	Endpoint	Value	Reference
Sarafloxacin	Antibiotic	S. fontinalis	Fish	48-h LC_{50}	>100	Wilford (1966)
		I. punctatus	Fish	48-h LC_{50}	>100	Wilford (1966)
		L. macrochirus	Fish	48-h LC_{50}	>100	Wilford (1966)
		S. namaycush	Fish	48-h LC_{50}	>100	Wilford (1966)
Spiramycin	Antibiotic	S. capricornutum	Algae	72-h EC_{50}	16	Hulton et al. (1999)
		M. aeruginosa	Algae	EC_{50}	0.005	Halling-Sorensen (2000)
		S. capricornutum	Algae	EC_{50}	2.3	Halling-Sorensen (2000)
Stavudine	Antibiotic (retroviral)	Daphnia spp.	Invert.	LC_{50}	>980	FDA-CDER (1996)
Streptomycin	Antibiotic	6 blue-green algae sp.	Algae	EC_{50}	0.09–0.86	Harrass et al. (1985)
Sulfachlorpyridazine	Antibiotic	V. fischeri	Bacteria	5-min LC_{50}	53.7	Kim et al. (2007)
		V. fischeri	Bacteria	15-min EC_{50}	26.4	Kim et al. (2007)
		D. magna	Invert.	48-h LC_{50}	357.3	Kim et al. (2007)
		D. magna	Invert.	96-h EC_{50}	233.5	Kim et al. (2007)
		O. latipes	Fish	48-h LC_{50}	589.3	Kim et al. (2007)
		O. latipes	Fish	96-h LC_{50}	535.7	Kim et al. (2007)
		L. minor	Plant	48-h EC_{50}	2.33	Pro et al. (2003)
Sulfadiazine	Antibiotic	D. magna	Invert.	48-h EC_{50}	221	Yamashita et al. (2006)
		D. magna	Invert.	24-h LOEC	150	Yamashita et al. (2006)
		D. magna	Invert.	48-h EC_{50}	212	De Liguoro et al. (2009)
		M. aeruginosa	Algae	72-h EC_{50}	0.135	Hulton et al. (1999)
		S. capricornutum	Algae	72-h EC_{50}	7.8	Hulton et al. (1999)
		S. capricornutum	Algae	EC_{50}, NOEC	2.19, <1.00	Costanzo et al. (2005)
Sulfadimethoxine	Antibiotic	V. fisheri	Bacteria	5-min LC_{50}	>500	Kim et al. (2007)
		V. fisheri	Bacteria	15-min EC_{50}	>500	Kim et al. (2007)
		D. magna	Invert.	48-h LC_{50}	248	Kim et al. (2007)
		D. magna	Invert.	96-h EC_{50}	204.5	Kim et al. (2007)

(continued)

Table 1 (continued)

Compound	Category	Species	Trophic group	Endpoint/Duration[a]	Value (mg/L)	References
		D. magna	Invert.	48-h EC$_{50}$	270	De Liguoro et al. (2009)
		D. magna	Invert.	24-h EC$_{50}$	639.8	Park and Cjoi (2008)
		M. macrocopa	Invert.	24, 48-h EC$_{50}$	296.6, 183.9	Park and Cjoi (2008)
		O. latipes	Fish	48-h LC$_{50}$	>100	Kim et al. (2007)
		O. latipes	Fish	96-h LC$_{50}$	>100	Kim et al. (2007)
		S. capricornutum	Algae	EC$_{50}$, NOEC	2.3, 0.529	Costanzo et al. (2005)
		C. vulgaris	Algae	EC$_{50}$, NOEC	11.2, <20.3	Costanzo et al. (2005)
Sulfamerazine	Antibiotic	O. mykiss	Fish	48-h LC$_{50}$	>100	Wilford (1966)
		S. trutta	Fish	48-h LC$_{50}$	>100	Wilford (1966)
		S. fontinalis	Fish	48-h LC$_{50}$	>100	Wilford (1966)
		I. punctatus	Fish	48-h LC$_{50}$	>100	Wilford (1966)
		L. macrochirus	Fish	48-h LC$_{50}$	>100	Wilford (1966)
		S. namaycush	Fish	48-h LC$_{50}$	>100	Wilford (1966)
Sulfamethazine	Antibiotic	V. fisheri	Bacteria	5-min LC$_{50}$	303	Kim et al. (2007)
		V. fisheri	Bacteria	15-min EC$_{50}$	344.7	Kim et al. (2007)
		D. magna	Invert.	48-h LC$_{50}$	174.4	Kim et al. (2007)
		D. magna	Invert.	96 EC$_{20}$	158.8	Kim et al. (2007)
		D. magna	Invert.	48-h EC$_{50}$	202	De Liguoro et al. (2009)
		D. magna	Invert.	24, 48-h EC$_{50}$	506.3, 215.9	Park and Cjoi (2008)
		M. macrocopa	Invert.	24, 48-h EC$_{50}$	310.9, 110.7	Park and Cjoi (2008)
		O. latipes	Fish	48-h LC$_{50}$	>100	Kim et al. (2007)
		O. latipes	Fish	96-h LC$_{50}$	>100	Kim et al. (2007)
		O. mykiss	Fish	48-h LC$_{50}$	>100	Wilford (1966)
		S. trutta	Fish	48-h LC$_{50}$	>100	Wilford (1966)
		S. fontinalis	Fish	48-h LC$_{50}$	>100	Wilford (1966)
		I. punctatus	Fish	48-h LC$_{50}$	>100	Wilford (1966)
		L. macrochirus	Fish	48-h LC$_{50}$	>100	Wilford (1966)
		S. namaycush	Fish	48-h LC$_{50}$	>100	Wilford (1966)

Compound	Class	Species	Endpoint	Value	Reference
Sulfamethoxazole	Antibiotic	V. fisheri	Bacteria	5-min LC$_{50}$	74.2	Kim et al. (2007)
		V. fisheri	Bacteria	15-min EC$_{50}$	78.1	Kim et al. (2007)
		V. fisheri	Bacteria	5-min EC$_{50}$	>84	Ferrari et al. (2003)
		D. magna	Invert.	48-h LC$_{50}$	189.2	Kim et al. (2007)
		D. magna	Invert.	96-h EC$_{50}$	177.3	Kim et al. (2007)
		D. magna	Invert.	48-h LC$_{50}$	177.3	Ferrari et al. (2003)
		D. magna	Invert.	48-h EC$_{50}$	202	De Liguoro et al. (2009)
		D. magna	Invert.	48-h EC$_{50}$	123.1	Park and Cjoi (2008)
		D. magna	Invert.	24-h EC$_{50}$	25.2	Isidori et al. (2005b)
		C. dubia	Invert.	24-h EC$_{50}$	15.51	Isidori et al. (2005b)
		M. macrocopa	Invert.	24, 48-h EC$_{50}$	84.9, 70.4	Park and Cjoi (2008)
		T. platyurus	Invert.	24-h LC$_{50}$	35.36	Isidori et al. (2005b)
		B. calyciflorus	Invert.	24-h LC$_{50}$	26.27	Isidori et al. (2005b)
		H. attenuata	Invert.	96-h LC$_{50}$, LOEC, NOEC	>100, 10, 5	Quinn et al. (2008)
		O. latipes	Fish	48-h LC$_{50}$	>750	Kim et al. (2007)
		O. latipes	Fish	96-h LC$_{50}$	>562	Kim et al. (2007)
		X. laevis	Amphib.	96-h EC$_{10}$	>100	Richards and Cole (2006)
		P. subcapitata	Algae	72-h EC$_{50}$	0.52	Isidori et al. (2005b)
		S. capricornutum	Algae	EC$_{50}$, NOEC	1.53, 0.614	Costanzo et al. (2005)
		C. meneghin-iana	Algae	96-h LC$_{50}$	2.4	Ferrari et al. (2003)
		S. leopolensis	Algae	96-h LOEC	0.006	Ferrari et al. (2003)
		P. subcapitata	Algae	96-h LOEC	0.09	Ferrari et al. (2003)
Sulfapyridine	Antibiotic	H. attenuata	Invert	96-h LC$_{50}$, EC$_{50}$, LOEC, NOEC	>100, 21.61, 5, 1	Quinn et al. (2008)
Sulfathaizole	Antibiotic	V. fisheri	Bacteria	5-min LC$_{50}$	1,000	Kuhn et al. (1989)
		V. fisheri	Bacteria	15-min EC$_{50}$	1,000	Kuhn et al. (1989)
		D. magna	Invert.	48-h LC$_{50}$	149.3	Kuhn et al. (1989)
		D. magna	Invert.	96 EC$_{50}$	85.4	Kuhn et al. (1989)
		D. magna	Invert.	24-h EC$_{50}$	616.7	Park and Cjoi (2008)

(continued)

Table 1 (continued)

Compound	Category	Species	Trophic group	Endpoint/Duration[a]	Value (mg/L)	References
Sulfisoxazole	Antibiotic	M. macrocopa	Invert.	24, 48-h EC$_{50}$	430.1, 391.1	Park and Cjoi (2008)
		O. latipes	Fish	48-h LC$_{50}$	>500	Kim et al. (2007)
		O. latipes	Fish	96-h LC$_{50}$	>500	Kim et al. (2007)
		O. mykiss	Fish	48-h LC$_{50}$	>100	Wilford (1966)
		S. trutta	Fish	48-h LC$_{50}$	>100	Wilford (1966)
		S. fontinalis	Fish	48-h LC$_{50}$	>100	Wilford (1966)
		I. punctatus	Fish	48-h LC$_{50}$	>100	Wilford (1966)
		L. macrochirus	Fish	48-h LC$_{50}$	>100	Wilford (1966)
		S. namaycush	Fish	48-h LC$_{50}$	>100	Wilford (1966)
Tetracycline	Antibiotic	D. magna	Invert.	48-h NOEC	340	Wollenberger et al. (2000)
		S. namaycush	Fish	24-h LC$_{50}$	220	Marking et al. (1988)
		M. saxatilis	Fish	24/48/96-h LC$_{50}$	>182	Welborn (1969)
		M. aeruginosa	Algae	EC$_{50}$	0.09	Halling-Sorensen (2000)
		S. capricornutum	Algae	EC$_{50}$	2.2	Halling-Sorensen (2000)
		Synechocystis sp.	Algae	120-h LOEC	0.01	Pomati et al. (2004)
Trimethoprim	Antibiotic	V. fischeri	Bacteria	5-min LC$_{50}$	165.1	Kim et al. (2007)
		V. fischeri	Bacteria	15-min EC$_{50}$	176.7	Kim et al. (2007)
		D. magna	Invert.	48-h LC$_{50}$	167.4	Kim et al. (2007)
		D. magna	Invert.	96-h EC$_{50}$	120.7	Kim et al. (2007)
		D. magna	Invert.	48-h LC$_{50}$	123	Robinson et al. (2005)
		D. magna	Invert.	48-h EC$_{50}$	149	De Liguoro et al. (2009)
		D. magna	Invert.	24, 48-h EC$_{50}$	155.6, 92	De Liguoro et al. (2009)
		M. macrocopa	Invert.	24, 48-h EC$_{50}$	144.8, 54.8	De Liguoro et al. (2009)
		H. attenuata	Invert.	96-h LC$_{50}$, LOEC, NOEC	>100,000	Quinn et al. (2008)
		O. latipes	Fish	48-h LC$_{50}$	>100	Kim et al. (2007)
		O. latipes	Fish	96-h LC$_{50}$	>100	Kim et al. (2007)
		X. laevis	Amphib.	96-h EC$_{50}$	>100	Richards and Cole (2006)
		M. aeruginosa	Algae	72-h EC$_{50}$	112,000	Hulton et al. (1999)

Compound	Class	Species	Type	Endpoint	Value	Reference
		S. capricornutum	Algae	72-h EC_{50}	130,000	Hulton et al. (1999)
		S. capricornutum	Algae	EC_{50}, NOEC	80,300, 25,500	Costanzo et al. (2005)
Tylosin	Antibiotic	*D. magna*	Invert.	24-h LOEC, 48-h EC_{50}	700, 680	Wollenberger et al. (2000)
		M. aeruginosa	Algae	EC_{50}	0.034	Halling-Sorensen (2000)
		S. capricornutum	Algae	EC_{50}	1.38	Halling-Sorensen (2000)
		S. capricornutum	Algae	EC_{50}, NOEC	0.411, 0.206	Costanzo et al. (2005)
Zalcitabine	Antibiotic (Retroviral)	*Daphnia* spp.	Invert.	EC_{50}	>1790	FDA-CDER (1996)
Atropine sulfate	Anticholinergic	*D. magna*	Invert.	24-h EC_{50}	258	Lilius et al. (1994)
		D. magna	Invert.	24-h EC_{50}	356	Calleja et al. (1994)
		S. proboscideus	Invert.	24-h LC_{50}	661	Calleja et al. (1994)
		B. calyciflorus	Invert.	24-h LC_{50}	334	Calleja et al. (1994)
Warfarin	Anticoagulant	*D. magna*	Invert.	24-h EC_{50}	89	Lilius et al. (1994)
		D. magna	Invert.	24-h EC_{50}	475	Calleja et al. (1994)
		S. proboscideus	Invert.	24-h LC_{50}	342	Calleja et al. (1994)
		B. calyciflorus	Invert.	24-h LC_{50}	444	Calleja et al. (1994)
Amitriptyline	Antidepressant	*D. magna*	Invert.	24-h EC_{50}	1.2	Lilius et al. (1994)
		D. magna	Invert.	24-h EC_{50}	5.55	Calleja et al. (1994)
		S. proboscideus	Invert.	24-h LC_{50}	0.78	Calleja et al. (1994)
		B. calyciflorus	Invert.	24-h LC_{50}	0.80	Calleja et al. (1994)
Citalopram	Antidepressant	*D. magna*	Invert.	48-h EC_{50}	20	Christensen et al. (2007)
		Ceriodaphnia dubia	Invert.	48-h LC_{50}	3.9	Henry et al. (2004)
		P. subcapitata	Algae	48-h EC_{50}	1.6	Christensen et al. (2007)
Fluoxetine	Antidepressant	*Daphnia* spp.	Invert.	EC_{50}	0.94	FDA-CDER (1996)
		D. magna	Invert.	48-h EC_{50}	6.4	Christensen et al. (2007)
		D. magna	Invert.	48-h LC_{50}	0.82	Brooks et al. (2003a)
		D. magna	Invert.	48-h LC_{50}	0.82	Brooks et al. (2003b)
		C. dubia	Invert.	48-h LC_{50}	0.51	Henry et al. (2004)
		C. dubia	Invert.	48-h LC_{50}, LOEC, NOEC	0.234, 0.112, 0.056	Brooks et al. (2003a)
		C. dubia	Invert.	48-h LC_{50}	0.234	Brooks et al. (2003b)
		O. mykiss	Fish	LC_{50}	2.0	FDA-CDER (1996)
		G. affinis	Fish	7-day LC_{50}	0.55	Henry and Black (2008)

(continued)

Table 1 (continued)

Compound	Category	Species	Trophic group	Endpoint/Duration[a]	Value (mg/L)	References
		P. promelas	Fish	48-h LC$_{50}$	0.705	Brooks et al. (2003a)
		P. promelas	Fish	48-h LC$_{50}$	0.705	Brooks et al. (2003b)
		X. laevis	Amphib.	96-h LC$_{10}$	7.1	Richards and Cole (2006)
		X. laevis	Amphib.	96-h EC$_{10}$	3	Richards and Cole (2006)
		Unspecified green algae	Algae	EC$_{50}$	0.031	FDA-CDER (1996)
		S. striatinum	Algae	4-h LOEC	1.55	Fong et al. (1998)
		P. subcapitata	Algae	48-h EC$_{50}$	0.027	Christensen et al. (2007)
		P. subcapitata	Algae	120-h EC$_{50}$, LOEC	0.024, 0.0136	Brooks et al. (2003a)
		P. subcapitata	Algae	120-h EC$_{50}$	0.039	Brooks et al. (2003b)
		P. subcapitata	Algae	96-h EC$_{50}$	0.04499	Johnson et al. (2007)
		S. acutus	Algae	96-h EC$_{50}$	0.09123	Johnson et al. (2007)
		S. quadricauda	Algae	96-h EC$_{50}$	0.21298	Johnson et al. (2007)
		C. vulgaris	Algae	96-h EC$_{50}$	4.33925	Johnson et al. (2007)
Fluvoxamine	Antidepressant	D. magna	Invert.	48-h EC$_{50}$	13	Christensen et al. (2007)
		Unspecified algae	Algae	MIC	63	FDA-CDER (1996)
		S. striatinum	Algae	4-h LOEC	0.003	Fong et al. (1998)
		P. subcapitata	Algae	48-h EC$_{50}$	0.062	Christensen et al. (2007)
		P. subcapitata	Algae	96-h EC$_{50}$	4.00	Johnson et al. (2007)
		S. acutus	Algae	96-h EC$_{50}$	3.62	Johnson et al. (2007)
		S. quadricauda	Algae	96-h-h EC$_{50}$	3.56	Johnson et al. (2007)
		C. vulgaris	Algae	96-h EC$_{50}$	10.21	Johnson et al. (2007)
Lithium sulfate	Antidepressant	D. magna	Invert.	24-h EC$_{50}$	197	Lilius et al. (1994)
		D. magna	Invert.	24-h EC$_{50}$	33.1	Calleja et al. (1994)
		S. proboscideus	Invert.	24-h LC$_{50}$	112	Calleja et al. (1994)
		B. calyciflorus	Invert.	24-h LC$_{50}$	712	Calleja et al. (1994)
Nefazodone	Antidepressant	Daphnia spp.	Invert.	EC$_{50}$	7	FDA-CDER (1996)

Drug	Category	Species	Group	Endpoint	Value	Reference
Paroxetine	Antidepressant	Daphnia spp.	Invert.	EC_{50}	3.0	FDA-CDER (1996)
		D. magna	Invert.	48-h EC_{50}	2.5	Cunningham et al. (2004)
		L. macrochirus	Fish	LC_{50}	2.0	FDA-CDER (1996)
		X. laevis	Amphib.	96-h LC_{10}	4.4	Richards and Cole (2006)
		X. laevis	Amphib.	96-h EC_{10}	3.6	Richards and Cole (2006)
		P. subcapitata	Algae	48-h EC_{50}	0.140	Christensen et al. (2007)
		S. striatinum	Algae	4-h LOEC	3.29	Fong et al. (1998)
Sertraline	Antidepressant	D. magna	Invert.	48-h EC_{50}	0.92	Christensen et al. (2007)
		D. magna	Invert.	24-h EC_{50}	3.1	Heckmann et al. (2007)
		D. magna	Invert.	48-h EC_{50}, LOEC, NOEC	1.3, 0.18, 0.10	Heckmann et al. (2007)
		C. dubia	Invert.	48-h LC_{50}	0.12	Henry et al. (2004)
		T. platyurus	Invert.	24-h LC_{50}, LOEC, NOEC	> 0.6, 0.6, 0.4	Heckmann et al. (2007)
		O. mykiss	Fish	96-h LC_{50}, LOEC, NOEC	0.38, 0.32, 0.1	Heckmann et al. (2007)
		X. laevis	Amphib.	96-h LC_{10}	3.6	Richards and Cole (2006)
		X. laevis	Amphib.	96-h EC_{10}	3	Richards and Cole (2006)
		P. subcapitata	Algae	48-h EC_{50}	0.04	Christensen et al. (2007)
		P. subcapitata	Algae	72-h EC_{50}, LOEC, NOEC	0.14, 0.075, 0.05	Heckmann et al. (2007)
		P. subcapitata	Algae	96-h EC_{50}	0.012	Johnson et al. (2007)
		S. acutus	Algae	96-h EC_{50}	0.099	Johnson et al. (2007)
		S. quadricauda	Algae	96-h EC_{50}	0.317	Johnson et al. (2007)
		C. vulgaris	Algae	96-h EC_{50}	0.764	Johnson et al. (2007)
Acarbose	Antidiabetic	Daphnia spp.	Invert.	EC_{50}	>1,000	FDA-CDER (1996)
		Unspecified fish	Fish	EC_{50}	>1,000	FDA-CDER (1996)
Metformin	Antidiabetic	Daphnia spp.	Invert.	EC_{50}	130	FDA-CDER (1996)
		D. magna	Invert.	48-h EC_{50}	64	Cleuvers (2003)
		L. macrochirus	Fish	LC_{50}	>982	FDA-CDER (1996)
		D. subspicatus	Algae	72-h EC_{50}	>320	Cleuvers (2003)
Ondansetron	Antiemetic	Daphnia spp.	Invert.	EC_{50}	28	FDA-CDER (1996)

(continued)

Table 1 (continued)

Compound	Category	Species	Trophic group	Endpoint/Duration[a]	Value (mg/L)	References
Carbamazepine	Antiepilectic	Biofilm		EC_{50}	0.01	Lin et al. (2009)
		V. fischeri	Bacteria	30-min EC_{50}	>81	Ferrari et al. (2003)
		V. fischeri	Bacteria	5-min EC_{50}	0.37 (µM)	Jos et al. (2003)
		V. fischeri	Bacteria	15-min EC_{50}	0.332 (µM)	Jos et al. (2003)
		V. fischeri	Bacteria	60-min EC_{50}	0.272 (µM)	Jos et al. (2003)
		V. fischeri	Bacteria	5-min EC_{50}	52.2	Kuhn et al. (1989)
		V. fischeri	Bacteria	15-min EC_{50}	52.2	Kuhn et al. (1989)
		V. fischeri	Bacteria	5-min EC_{50}	>81	Ferrari et al. (2003)
		C. vulgaris	Invert.	24-h EC_{50}	469.5 (µM)	Jos et al. (2003)
		C. vulgaris	Invert.	48-h EC_{50}	155 (µM)	Jos et al. (2003)
		C. dubia	Invert.	48-h EC_{50}	77.7	Ferrari et al. (2003)
		C. dubia	Invert.	48-h EC_{50}	77.7	Ferrari et al. (2003)
		D. magna	Invert.	48-h EC_{50}	>100	Cleuvers (2003)
		D. magna	Invert.	24-h EC_{50}	475 (µM)	Jos et al. (2003)
		D. magna	Invert.	48-h EC_{50}	414 (µM)	Jos et al. (2003)
		D. magna	Invert.	48-h LC_{50}	>100	Kim et al. (2007)
		D. magna	Invert.	96 EC_{50}	76.3	Kim et al. (2007)
		D. magna	Invert.	48-h EC_{50}	>13.8	Ferrari et al. (2003)
		H. attenuata	Invert.	96-h LC_{50}, EC_{50}, LOEC, NOEC	29.4, 15.52, 5, 1	Quinn et al. (2008)
		H. attenuata	Invert.	96-h feeding behavior	3.76	Quinn et al. (2008)
		T. platyurus	Invert.	24-h LC_{50}	>100	Kim et al. (2009)
		O. latipes	Fish	48-h LC_{50}	35.4	Kim et al. (2007)
		O. latipes	Fish	96-h LC_{50}	45.87	Kim et al. (2009)
		O. mykiss (juvenile)	Fish	96-h LC_{50}	19.9	Li et al. (2011)
		D. rerio (embryo)	Fish	72-h EC_{50}	86.5	van den Brandof and Montforts (2010)
		P. subcapitata	Algae	96-h LOEC, NOEC	>100, >100	Ferrari et al. (2003)
		P. subscapitata	Algae	96-h LOEC	>100	Ferrari et al. (2003)

Drug	Class	Species	Group	Endpoint	Value	Reference
		D. subspicatus	Algae	72-h EC_{50}	74	Cleuvers (2003)
		C. meneghiniana	Algae	96-h LC_{50}	31.6	Ferrari et al. (2003)
		C. meneghiniana	Algae	96-h EC_{50}	10	Ferrari et al. (2003)
		S. leopolensis	Algae	96-h EC_{50}	17	Ferrari et al. (2003)
		X. laevis	Amphib.	96-h EC_{10}	>100	Richards and Cole (2006)
Gabapentin	Antiepileptic	*Daphnia* spp.	Invert.	EC_{50}	>100	FDA-CDER (1996)
Faslodex	Antiestrogen	*D. magna*	Invert.	96-h EC_{50}	0.129	Clubbs and Brooks (2007)
Diphenhydramine	Antihistamine	*D. magna*	Invert.	48-h LC_{50}	0.374	Berninger et al. (2011)
		P. promelas	Fish	48-h LC_{50}	59.28	Berninger et al. (2011)
		P. promelas	Fish	48-h LC_{50}	2.09	Berninger et al. (2011)
Cetirizine	Antihistaminic	*Daphnia* spp.	Invert.	EC_{50}	330	FDA-CDER (1996)
Amlodipine	Antihypertensive	*H. vulgaris*	Invert.	7-day LC_{50}	>100	Pascoe et al. (2003)
Atenolol	Antihypertensive	*H. vulgaris*	Invert.	7-day LC_{50}	>100	Pascoe et al. (2003)
		D. magna	Invert.	48-h EC_{50}	313	Cleuvers (2005)
		C. dubia	Invert.	48-h EC_{50}	38	Fraysse and Garric (2005)
		T. platyurus	Invert.	24-h LC_{50}	>100	Kim et al. (2009)
		O. latipes	Fish	96-h LC_{50}	>100	Kim et al. (2009)
		D. subspicatus	Algae	72-h EC_{50}	620	Cleuvers (2005)
		P. subcapitata	Algae	72-h NOEC	128.8	Kuster et al. (2009)
Captopril	Antihypertensive	*D. magna*	Invert.	48-h EC_{50}	>100	Cleuvers (2003)
		D. subspicatus	Algae	72-h EC_{50}	168	Cleuvers (2003)
Carvedilol	Antihypertensive	*Daphnia* spp.	Invert.	EC_{50}	>3	FDA-CDER (1996)
		Unspecified fish	Fish	LC_{50}	1	FDA-CDER (1996)
Diltiazem	Antihypertensive	*V. fischeri*	Bacteria	5-min LC_{50}	407	Kim et al. (2007)
		V. fischeri	Bacteria	15-min EC_{50}	263	Kim et al. (2007)
		D. magna	Invert.	48-h LC_{50}	28	Kim et al. (2007)
		D. magna	Invert.	96-h EC_{50}	8.2	Kim et al. (2007)
		O. latipes	Fish	48-h LC_{50}	25.6	Kim et al. (2007)
		O. latipes	Fish	96-h LC_{50}	15	Kim et al. (2007)

(continued)

Table 1 (continued)

Compound	Category	Species	Trophic group	Endpoint/Duration[a]	Value (mg/L)	References
Losartan K	Antihypertensive	Daphnia spp.	Invert.	EC$_{50}$	331	FDA-CDER (1996)
		O. mykiss	Fish	LC$_{50}$	>929	FDA-CDER (1996)
		P. promelas	Fish	LC$_{50}$	>1,000	FDA-CDER (1996)
		Unspecified green algae	Algae	MIC	245	FDA-CDER (1996)
		Unspecified blue-green algae	Algae	MIC	949	FDA-CDER (1996)
Metoprolol	Antihypertensive	D. magna	Invert.	48-h EC$_{50}$	>100	Cleuvers (2003)
		D. magna	Invert.	48-h EC$_{50}$	>100	Cleuvers (2003)
		D. magna	Invert.	48-h LC$_{50}$	63.9	Huggett et al. (2002)
		D. magna	Invert.	48-h LC$_{50}$	438	Cleuvers (2005)
		C. dubia	Invert.	48-h LC$_{50}$	8.8	Huggett et al. (2002)
		H. azteca	Invert.	48-h LC$_{50}$	>100	Huggett et al. (2002)
		O. latipes	Fish	48-h LC$_{50}$	>100	Huggett et al. (2002)
		D. rerio (embryo)	Fish	72-h EC$_{50}$	31.0	van den Brandof and Montforts (2010)
		D. subspicatus	Algae	48-h EC$_{50}$	7.9	Brooks et al. (2003a)
		D. subspicatus	Algae	48-h EC$_{50}$	7.3	Cleuvers (2003)
		D. subspicatus	Algae	72-h EC$_{50}$	7.3	Cleuvers (2003)
		L. minor	Plant	7-day EC$_{50}$	>320	Cleuvers (2003)
Moexipril	Antihypertensive	Daphnia spp.	Invert.	EC$_{50}$	800	FDA-CDER (1996)
Moexiprilat	Antihypertensive	Daphnia spp.	Invert.	EC$_{50}$	>1,000	FDA-CDER (1996)
Nisoldipine	Antihypertensive	Daphnia spp.	Invert.	EC$_{50}$	33	FDA-CDER (1996)
		Unspecified fish	Fish	EC$_{50}$	3	FDA-CDER (1996)
Perindopril Erbumine	Antihypertensive	Daphnia spp.	Invert.	EC$_{50}$	>1,000	FDA-CDER (1996)
		L. macrochirus	Fish	LC$_{50}$	>990	FDA-CDER (1996)
Propranolol	Antihypertensive	D. magna	Invert.	48-h LC$_{50}$	3.1–17.7	Lilius et al. (1994)
		D. magna	Invert.	48-h EC$_{50}$	7.5	Cleuvers (2003)
		D. magna	Invert.	24-h EC$_{50}$	2.7	Lilius et al. (1994)

Drug	Therapeutic class	Species	Organism	Endpoint	Value	Reference
		D. magna	Invert.	48-h EC$_{50}$	1.6	Huggett et al. (2002)
		D. magna	Invert.	48-h EC$_{50}$	1.67	Stanley et al. (2006)
		D. magna	Invert.	48-h EC$_{50}$	7.7	Cleuvers (2005)
		C. dubia	Invert.	48-h LC$_{50}$	0.8	Huggett et al. (2002)
		T. platyurus	Invert.	24-h LC$_{50}$	10.31	Kim et al. (2009)
		H. azteca	Invert.	48-h LC$_{50}$	29.8	Huggett et al. (2002)
		O. latipes	Fish	96-h LC$_{50}$	11.4	Kim et al. (2009)
		O. latipes	Fish	48-h LC$_{50}$	24.3	Huggett et al. (2002)
		P. promelas	Fish	48-h LC$_{50}$	1.21	Huggett et al. (2002)
		D. subspicatus	Algae	72-h EC$_{50}$	5.8	Cleuvers (2003)
		D. subspicatus	Algae	48-h EC$_{50}$	0.7	Cleuvers (2005)
		S. leopolensis	Algae	96-h EC$_{50}$	0.35	Ferrari et al. (2003)
		P. subcapitata	Algae	96-h EC$_{50}$	0.5	Ferrari et al. (2003)
		L. minor	Plant	7-day growth	114	Cleuvers (2003)
R-Propranolol	Antihypertensive	S. proboscideus	Invert.	24-h LC$_{50}$	1.87	Calleja et al. (1994)
		D. magna	Invert.	24-h EC$_{50}$	15.87	Calleja et al. (1994)
		D. magna	Invert.	48-h EC$_{50}$	1.57	Stanley et al. (2006)
		B. calyciflorus	Invert.	24-h LC$_{50}$	2.59	Calleja et al. (1994)
		P. promelas	Fish	48-h LC$_{50}$	1.69	Stanley et al. (2006)
S-Propranolol	Antihypertensive	D. magna	Invert.	EC$_{50}$	1.4	Stanley et al. (2006)
		P. promelas	Fish	LC$_{50}$	1.42	Stanley et al. (2006)
Spirapri	Antihypertensive	Daphnia spp.	Invert.	EC$_{50}$	>930	FDA-CDER (1996)
		L. macrochirus	Fish	LC$_{50}$	>970	FDA-CDER (1996)
Sumatriptan succinate	Anti-migraine	Daphnia spp.	Invert.	EC$_{50}$	290	FDA-CDER (1996)
Cladribine	Antineoplastic	Daphnia spp.	Invert.	EC$_{50}$	233	FDA-CDER (1996)
Cyclophosphamide	Antineoplastic	P. subcapitata	Algae	72-h EC$_{50}$, NOEC	>100, >100	Grung et al. (2008)
Methotrexate	Antineoplastic	D. magna	Invert.	EC$_{50}$	>1,000	Henschel et al. (1997)
		T. pyriformis	Invert.	48-h EC$_{50}$	45	Henschel et al. (1997)
		B. cayciflorus	Invert.	24-h LC$_{50}$	0.97	DellaGreca et al. (2007)
		B. rerio	Fish	48-h EC$_{50}$	850	Henschel et al. (1997)
		S. subspicatus	Algae	72-h EC$_{50}$	260	Henschel et al. (1997)

(continued)

Table 1 (continued)

Compound	Category	Species	Trophic group	Endpoint/Duration[a]	Value (mg/L)	References
Paclitaxel	Antineoplastic	Daphnia spp.	Invert.	LC_{50}	>0.74	FDA-CDER (1996)
Tamoxifen	Antineoplastic	D. magna	Invert.	24-h EC_{50}	1.53	DellaGreca et al. (2007)
		T. platyurus	Invert.	24-h LC_{50}	0.40	DellaGreca et al. (2007)
Thiotepa	Antineoplastic	Daphnia spp.	Invert.	EC_{50}	546	FDA-CDER (1996)
Risperidone	Antipsychotic	Daphnia spp.	Invert.	EC_{50}	6.0	FDA-CDER (1996)
		L. macrochirus	Fish	LC_{50}	6.0	FDA-CDER (1996)
Thioridazine	Antipsychotic	D. magna	Invert.	24-h EC_{50}	0.69	Lilius et al. (1994)
		D. magna	Invert.	24-h EC_{50}	4.56	Calleja et al. (1994)
		S. proboscideus	Invert.	24-h LC_{50}	0.33	Calleja et al. (1994)
		B. calyciflorus	Invert.	24-h LC_{50}	0.30	Calleja et al. (1994)
Cimetidine	Antiulcerative	V. fischeri	Bacteria	5-month LC_{50}	458.9	Kim et al. (2007)
		D. magna	Invert.	48-h LC_{50}	379.7	Kim et al. (2007)
		D. magna	Invert.	96-h EC_{50}	271.3	Kim et al. (2007)
		Daphnia spp.	Invert.	EC_{50}	740	FDA-CDER (1996)
		L. macrochirus	Fish	LC_{50}	>1,000	FDA-CDER (1996)
		O. latipes	Fish	96-h LC_{50}	>100	Kim et al. (2007)
Famotidine	Antiulcerative	D. magna	Invert.	EC_{50}	398	FDA-CDER (1996)
		P. promelas	Fish	LC_{50}	>680	FDA-CDER (1996)
Lansoprazole	Antiulcerative	Daphnia spp.	Invert.	EC_{50}	>22	FDA-CDER (1996)
		O. mykiss	Fish	LC_{50}	18	FDA-CDER (1996)
Omeprazole	Antiulcerative	Daphnia spp.	Invert.	EC_{50}	88	FDA-CDER (1996)
Ranitidine	Antiulcerative	Daphnia spp.	Invert.	EC_{50}	650	FDA-CDER (1996)
Flumazenil	Benzodiazepine antagonist	D. magna	Invert.	EC_{50}	>500	FDA-CDER (1996)
Alendronate sodium	Bone resorption inhibitor	Daphnia spp.	Invert.	LC_{50}	22	FDA-CDER (1996)
		P. promelas	Fish	LC_{50}	1,450	FDA-CDER (1996)
		O. mykiss	Fish	LC_{50}	>1,000	FDA-CDER (1996)
		Green algae	Algae	MIC	>0.5	FDA-CDER (1996)
Etidronic acid	Bone resorption inhibitor	D. magna	Invert.	48-h EC_{50}	527	Gledhill and Feijtel (1992)
		O. mykiss	Fish	96-h LC_{50}	200	Gledhill and Feijtel (1992)

		Species		Endpoint	Value	Reference
Tiludronate disodium	Bone resorption inhibitor	L. macrochirus	Fish	96-h LC$_{50}$	868	Gledhill and Feijtel (1992)
		I. punctatus	Fish	48-h LC$_{50}$	695	Gledhill and Feijtel (1992)
		P. subcapitata	Algae	96-h EC$_{50}$	1.3–13.2	Gledhill and Feijtel (1992)
		Unspecified algae	Algae	96-h EC$_{50}$	3.0	Gledhill and Feijtel (1992)
		D. magna	Invert.	24-h EC$_{50}$	562	Sanofi (1996)
		D. magna	Invert.	48-h EC$_{50}$	320	Sanofi (1996)
		M. aeruginosa	Algae	96-h EC$_{50}$	13.3	Sanofi (1996)
		P. subcapitata	Algae	96-h EC$_{50}$	36.6	Sanofi (1996)
Dorzolamide	Carbonic anhydrase inhibitor	D. magna	Invert.	EC$_{50}$	699	FDA-CDER (1996)
		P. promelas	Fish	LC$_{50}$	>1,000	FDA-CDER (1996)
Digoxin	Cardiac stimulant	H. vulgaris	Invert.	7-day LC$_{50}$	>100	Pascoe et al. (2003)
		D. magna	Invert.	24-h EC$_{50}$	24	Lilius et al. (1994)
Milrinone lactate	Cardiotonic	Daphnia spp.	Invert.	EC$_{50}$	414	FDA-CDER (1996)
Nicotine sulfate	Cholinergic agonist	P. promelas	Fish	96-h EC$_{50}$	13.8	Russom et al. (1997)
Nicotine	Cholinergic agonist	D. magna	Invert.	EC$_{50}$	3.0	FDA-CDER (1996)
		O. mykiss	Fish	LC$_{50}$	7.0	FDA-CDER (1996)
		P. promelas	Fish	LC$_{50}$	20.0	FDA-CDER (1996)
		L. macrochirus	Fish	LC$_{50}$	4.0	FDA-CDER (1996)
		Goldfish	Fish	LC$_{50}$	13	FDA-CDER (1996)
Amphetamine sulfate	CNS stimulant	D. magna	Invert.	24-h EC$_{50}$	270	Calleja et al. (1994)
		D. magna	Invert.	24-h EC$_{50}$	60	Lilius et al. (1994)
		S. proboscideus	Invert.	24-h LC$_{50}$	5	Calleja et al. (1994)
		B. calyciflorus	Invert.	24-h LC$_{50}$	4.90	Calleja et al. (1994)
		P. promelas	Fish	96-h EC$_{50}$	28.8	Russom et al. (1997)
Caffeine	CNS stimulant	Biofilm		EC$_{50}$	0.01	Lin et al. (2009)
		C. dubia	Invert.	48-h LC$_{50}$	50–60	Moore et al. (2008)
		D. magna	Invert.	24-h EC$_{50}$	684	Lilius et al. (1994)
		D. magna	Invert.	24-h EC$_{50}$	160	Calleja et al. (1994)
		S. proboscideus	Invert.	24-h LC$_{50}$	410	Calleja et al. (1994)
		Chironomus dilutus	Invert.	48-h LC$_{50}$	970–1520	Moore et al. (2008)
		P. promelas	Fish	96-h EC$_{50}$	151	Russom et al. (1997)
		B. calyciflorus	Invert.	24-h LC$_{50}$	4,661	Calleja et al. (1994)

(continued)

Table 1 (continued)

Compound	Category	Species	Trophic group	Endpoint/Duration[a]	Value (mg/L)	References
		P. promelas	Fish	48-h LC$_{50}$	80–120	Moore et al. (2008)
		X. laevis	Amphib.	96-h EC$_{50}$	74–158	Bantle et al. (1994)
		X. laevis	Amphib.	96-h EC$_{50}$	240–350	De Young et al. (1996)
Iopromide	Diagnostic aid	Daphnia spp.	Invert.	EC$_{50}$	>1,016	FDA-CDER (1996)
		D. magna	Invert.	24-h EC$_{50}$	>10,000	Schweinfurth et al. (1996)
		D. magna	Invert.	48-h EC$_{50}$	>1,000	Gyllenhammar et al. (2009)
		O. mykiss	Fish	LC$_{50}$	>962	FDA-CDER (1996)
		L. macrochirus	Fish	LC$_{50}$	>973	FDA-CDER (1996)
		D. rerio	Fish	96-h LC$_{50}$	>10,000	Vandenbergh et al. (2003)
		L. idus	Fish	48-h LC$_{50}$	>10,000	Vandenbergh et al. (2003)
		Unspecified fish	Fish	48-h EC$_{50}$	>10,000	Schweinfurth et al. (1996)
		Unspecified green algae	Algae	MIC	137	FDA-CDER (1996)
		S. subspicatus	Algae	72-h EC$_{50}$	>10,000	Vandenbergh et al. (2003)
Bendroflumethiazide	Diuretic	H. vulgaris	Invert.	7-day LC$_{50}$	>100	Pascoe et al. (2003)
Furosemide	Diuretic	Biofilm		EC$_{50}$	>0.01	Lin et al. (2009)
		H. vulgaris	Invert.	7-day LC$_{50}$	>100	Pascoe et al. (2003)
Aprotinin	Enzyme inhibitor	Daphnia spp.	Invert.	EC$_{50}$	>1,000	FDA-CDER (1996)
Diethylstilbestrol	Estrogen	D. magna	Invert.	LC$_{50}$	4.0	Coats et al. (1976)
		D. magna	Invert.	48-h LC$_{50}$	1.09	Zou and Fingerman (1997)
		D. magna	Invert.	48-h LC$_{50}$	1.2	Baldwin et al. (1995)
		C. pipiens	Invert.	48-h LC$_{50}$	4	Coats et al. (1976)
		Physa spp.	Invert.	LC$_{50}$	>10	Coats et al. (1976)
		G. affinis	Fish	48-h LC$_{50}$	>1	Coats et al. (1976)
		P. promelas	Fish	14-day LC$_{50}$	316	Panter et al. (1999)
		O. cardiacum	Algae	48-h LC$_{50}$	>100	Coats et al. (1976)
Ethinylestradiol	Estrogen	D. magna	Invert.	24-h EC$_{50}$	5.7	Kopf (1995)
		D. magna	Invert.	48-h EC$_{50}$	6.4	Schweinfurth et al. (1996)
		D. magna	Invert.	96-h EC$_{50}$	2.59	Clubbs and Brooks (2007)
		Unspecified algae	Algae	EC$_{50}$	0.84	Kopf (1995)
		O. mykiss	Fish	96-h EC$_{50}$	1.6	Schweinfurth et al. (1996)

		Species	Endpoint	Value	Reference
Bezafibrate	Lipid lowering	B. calyciflorus	24-h LC$_{50}$	60.91	Isidori et al. (2007)
		T. platyurus	24-h LC$_{50}$	39.69	Isidori et al. (2007)
		D. magna	24-h EC$_{50}$	100.08	Isidori et al. (2007)
		D. magna	48-h EC$_{50}$	240.40	Rosal et al. (2010)
		C. dubia	48-h EC$_{50}$	75.79	Isidori et al. (2007)
Clofibrate	Lipid lowering	D. magna	24-h EC$_{50}$	28.2	DellaGreca et al. (2007)
		D. magna	48-h EC$_{50}$	72	Cleuvers (2003)
		D. rerio	96-h LC$_{50}$	0.89	Isidori et al. (2007)
		Non-specific	EC$_{50}$	12	Kopf (1995)
		C. meneghiniana	96-h LOEC	>100	Ferrari et al. (2003)
		D. subspicatus	72-h EC$_{50}$	115	Cleuvers (2003)
		P. subcapitata	96-h LOEC	150	Ferrari et al. (2003)
		S. leopolensis	96-h LOEC	23.5	Ferrari et al. (2003)
		Unspecified algae	EC$_{50}$	12.0	Kopf (1995)
Clofibrinic acid	Lipid lowering	D. magna	EC$_{50}$	106	Henschel et al. (1997)
		D. magna	48-h EC$_{50}$	72	Cleuvers et al. (2003)
		D. magna	48-h EC$_{50}$	>200	Ferrari et al. (2003)
		D. magna	48-h EC$_{50}$	83.52	Rosal et al. (2010)
		C. dubia	48-h EC$_{50}$	>200	Ferrari et al. (2003)
		B. rerio	48-h EC$_{50}$	86.0	Henschel et al. (1997)
		S. subspicatus	72-h EC$_{50}$	89	Henschel et al. (1997)
Fenofibrate	Lipid lowering	B. calyciflorus	24-h LC$_{50}$	64.97	Isidori et al. (2007)
		D. magna	24-h LC$_{50}$	50.12	Isidori et al. (2007)
		D. magna	48-h EC$_{50}$	4.90	Rosal et al. (2010)
		P. subcapitata	72-h LOEC	6.25	Isidori et al. (2007)
Gemfibrozil	Lipid lowering	V. fischeri	5-min EC$_{50}$	258 (μM)	Zurita et al. (2007)
		C. vulgaris	48-h EC$_{50}$	195	Zurita et al. (2007)
		D. magna	48, 72-h EC$_{50}$	42.6, 30.0	Zurita et al. (2007), Zurita et al. (2007)
		D. magna	48-h LC$_{50}$	10.4	Han et al. (2006)
		D. magna	24-h EC$_{50}$	74.3	Isidori et al. (2007)
		D. magna	48-h EC$_{50}$	22.85	Rosal et al. (2010)
		B. calyciflorus	24-h LC$_{50}$	77.3	Isidori et al. (2007)

(continued)

Table 1 (continued)

Compound	Category	Species	Trophic group	Endpoint/Duration[a]	Value (mg/L)	References
		T. platyurus	Invert.	24-h LC_{50}	161.05	Isidori et al. (2007)
		T. battagliai	Invert.	48-h LC_{50}	127.6	Schmidt et al. (2011)
		C. vulgaris	Algae	72-h EC_{50}	150	Zurita et al. (2007)
		P. subcapitata	Algae	72-h EC_{50}	6.25	Isidori et al. (2007)
		S. costatum	Algae	72-h IC_{50}	56.3	Schmidt et al. (2011)
Cisapride	Peristaltic	Daphnia spp.	Invert.	EC_{50}	>1,000	FDA-CDER (1996)
		L. macrochirus	Fish	LC_{50}	>1,000	FDA-CDER (1996)
Porfirmer sodium	Photosensitiser	Daphnia spp.	Invert.	EC_{50}	>994	FDA-CDER (1996)
Amobarbital	Sedative	P. promelas	Fish	96-h EC_{50}	85.4	Russom et al. (1997)
Diazepam	Sedative	D. magna	Invert.	24-h EC_{50}	14.1	Calleja et al. (1994)
		D. magna	Invert.	24-h EC_{50}	4.3	Lilius et al. (1994)
		S. proboscideus	Invert.	24-h LC_{50}	103	Calleja et al. (1994)
		H. vulgaris	Invert.	7-day LC_{50}	>100	Pascoe et al. (2003)
		B. calyciflorus	Invert.	24-h LC_{50}	>10,000	Calleja et al. (1994)
Midazolam	Sedative	D. magna	Invert.	EC_{50}	0.2	FDA-CDER (1996)
Orphenadrine	Sedative	D. magna	Invert.	24-h EC_{50}	8.9	Lilius et al. (1994)
		D. magna	Invert.	24-h EC_{50}	10.6	Calleja et al. (1994)
		S. proboscideus	Invert.	24-h LC_{50}	4.3	Calleja et al. (1994)
		B. calcyflorus	Invert.	24-h LC_{50}	5.4	Calleja et al. (1994)
Pentobarbital	Sedative	P. promelas	Fish	96-h EC_{50}	49.5	Russom et al. (1997)
Phenobarbital	Sedative	D. magna	Invert.	24-h EC_{50}	1,463	Calleja et al. (1994)
		S. proboscideus	Invert.	24-h LC_{50}	1,212	Calleja et al. (1994)
		B. calyciflorus	Invert.	24-h LC_{50}	5,179	Calleja et al. (1994)
		P. promelas	Fish	96-h EC_{50}	484	Russom et al. (1997)
Secobarbital	Sedative	P. promelas	Fish	96-h EC_{50}	23.6	Russom et al. (1997)
Thiopental, sodium salt	Sedative	P. promelas	Fish	96-h EC_{50}	26.2	Russom et al. (1997)

Table modified from Webb 2001 and Santos et al. 2010

[a]The LC_{xx} value is the lethal concentration causing mortality in $xx\%$ of organisms. EC_{xx} relates to the concentration eliciting an effect in $xx\%$ of organisms (in Daphnia effect refers to immobilization and algae relates to effects on growth). MIC is the minimum inhibitory concentration, LOEC is the lowest observable effects concentration, NOEC is the no observable effect concentration and NEL is the no effects level

trophic groups have been subjected to acute toxicity testing of at least a single pharmaceutical compound; however, only limited data are available for benthic invertebrates, bivalves, amphibians, and aquatic plants and algae.

2.1 Analgesics

Analgesic drugs include nonnarcotic drugs, nonsteroidal anti-inflammatory drugs (NSAIDs), and opiod drugs. Thus far, 12 different analgesic compounds have been investigated for acute aquatic-species toxicity. Ibuprofen, an NSAID, is the most commonly studied analgesic compound, and its acute toxicity has been evaluated in at least ten studies. Seven different trophic groups have been examined; the only major group without acute data appears to be aquatic plants. As a group, phytoplankton were the most sensitive to ibuprofen toxicity, exhibiting 72–120-h EC_{50} (concentration affecting 50% of organisms tested) values between 1 and 315 mg/L, depending on species (Pomati et al. 2004; Lawrence et al. 2005). The acute toxicity of paracetamol (or acetaminophen), a non-narcotic pain relieving drug, has been studied extensively in invertebrates; EC_{50} values (immobilization) have ranged from 13 to 290 mg/L for 24-h exposures and 50–92 mg/L for 48-h exposures (Kuhn et al. (1989); Calleja et al. 1994; Henschel et al. 1997; Kim et al. 2007). The narcotic analgesic dextropropoxyphene appears to be the most acutely toxic of the analgesics studied, with 24-h EC_{50} values in *D. magna* of 14–19 mg/L (Calleja et al. 1994; Lilius et al. 1994). The acute toxicity to *D. magna* has been examined for all analgesics, and this allows for comparison of toxicity among compounds. Ranking analgesic compounds by acute toxicity from the most to least toxic shows the following distribution: dextropropoxyphene (opiod)>paracetamol (non-narcotic)>tramadol (non-narcotic)>ibuprofen (NSAID)>naproxen sodium (NSAID)>diclofenac (NSAID)>salicylic acid (NSAID). In general, invertebrates and phytoplankton were most sensitive to the acute toxicity of the analgesics, whereas bacteria, fish, and amphibians were relatively insensitive.

2.2 Anti-androgens

Three anti-androgenic compounds have been investigated for acute toxicity to a range of species and trophic groups. Bicalutamide is a nonsteroidal anti-androgen that functions by binding to androgen receptors thus preventing upregulation of androgen receptive genes (Furr and Tucker 1996). Bicalutamide is prescribed in combination for advanced prostate cancer, and as a monotherapy for early stage prostate cancer. LC_{50}s (lethal concentration causing mortality in 50% of organisms tested) have been determined for bicalutamide in *Daphnia* sp., green algae, and blue-green algae; the acute toxicity values were >5 mg/L for *D. magna* and 1 mg/L for algae (FDA-CDER 1996). Finasteride is an anti-androgen used in the treatment of benign prostatic hypertrophy (or BPH), and works by inhibiting the enzyme (5-alpha reductase) that converts testosterone to dihydrotestosterone. It induces acute effects to invertebrates (*D. magna*) and fish (*Oncorhynchus mykiss*) at similar concentrations (i.e., 20 mg/L; FDA-CDER 1996) as bicalutamide. Based on the limited number of acute studies conducted for anti-androgen compounds, these drugs pose little acute risk at their anticipated environmental concentrations.

2.3 Anti-arrhythmics

Both quinidine sulfate, which blocks fast inward sodium current (Williams 1958), and verapamil, a voltage-gated calcium channel blocker (Nayler and Poole-Wilson 1981), are acutely toxic to aquatic invertebrates (i.e., *D. magna*, *S. proboscideus*, and *B. calyciflorus*), the only trophic group studied to date. Quinidine sulfate is more toxic to *B. calicyflorus* (8.7 mg/L), whereas verapamil is more toxic to *D. magna* and *S. proboscideus* (55.5 and 6.24 mg/L, respectively) (Calleja et al. 1994); however, neither compound is likely to cause acute effects in the environment.

2.4 Anti-asthmatics

The bronchodilator theophylline is used to treat respiratory diseases such as asthma. The acute toxicity of this drug has been investigated in *D. magna*, *S. proboscideus*, and *B. calyciflorus*. *D. magna* immobilization is the most sensitive endpoint observed, with 24-h EC_{50} values of 155 mg/L in one study (Lilius et al. 1994) and 483 mg/L in another (Calleja et al. 1994); however, these concentrations are much greater than what has been observed in aquatic ecosystems. *B. calciflorus* was relatively insensitive, displaying 24-h LC_{50} values >3 g/L (Calleja et al. 1994). Another anti-asthmatic drug, the corticosteroid fluticasone propionate, is much more acutely toxic than theophylline to invertebrates, with an EC_{50} value of 0.55 mg/L for *Daphnia* spp. (FDA-CDER 1996). The acute toxicity of salmeterol (β2-adrenergic receptor agonist; Ullman and Svedmyr 1988) has also been investigated in *D. magna*, and appears to be less acutely toxic ($EC_{50}=20$ mg/L) than fluticasone propionate, but significantly more toxic than theophylline (FDA-CDER 1996).

2.5 Antibiotics

The acute toxicity of antibiotics in aquatic species is the most studied of all pharmaceutical chemical classes. For simplicity, antibiotics will be divided into the following classes: anitamebics, antibacterials, antimalarials, antiprotozoals, antiseptics, biocides, and retrovirals. The antibacterial compounds are the most thoroughly studied of pharmaceutical compound groups from an environmental point of view. There have been 40 different antibiotic compounds, representing different antibiotic classes, studied to date. In general, no antibiotic class (i.e., fluoroquinolone, sulfonamide, etc.) is appreciably more acutely toxic than any other class, although algae are the most sensitive trophic group to antibiotics. Algae and plants retain numerous target sites for toxic action that are well conserved from bacterial ancestry (Bach and Lichtenthaler 1983; Lichtenthaler et al. 1997a, b; McFadden and Roos 1999). For example, fluoroquinolone antibiotics inhibit DNA-gyrase in bacteria, and evidence exists that demonstrate plants contain similar DNA-gyrase enzymes (Thompson and Mosig 1985; Wall et al. 2004). Moreover, triclosan, a chlorophenol antibiotic, disrupts bacterial fatty acid synthesis (McMurry et al. 1998; Heath et al. 1999) that can be found in both bacteria and plants. Similarly, the tetracycline, macrolide, lincosamide, and pleuromutilin antibiotics inhibit transcription and translation in bacteria and plants (Manuell et al. 2004) and sulfonamides inhibits folate synthesis pathways in both plants and bacteria (Sweetman 2002; Basset et al. 2005; Brain et al. 2008a). For a full review of pharmaceutical effects on plants refer to Brain et al. (2008b) and Brain and Cedergreen (2008).

Acute toxicity thresholds of the antiseptic compounds, acriflavine, nitrofurazone, and loracarbef have been identified in numerous trophic groups; fish have been the group most frequently tested. All fish are equally sensitive to these three compounds. The recorded 24-h and 48-h LC_{50} values ranged between 13.5 and 48 mg/L (Wilford 1966; Hughes 1973). Nitrofurazone has also been studied in algae and invertebrates, and algae are more sensitive than invertebrates or fish (Macri and Sbardella 1984). Thimerosal (or merthiolate) is a preservative used in vaccines and has been investigated for acute toxicity in a wide variety of fish species. Thimerosal is most toxic to cold-water salmonids (*O. mykiss*, *Salvelinus namaycush*, and *Salmo trutta*), with 48-h LC_{50} values ranging between 2.1 and 54 mg/L (Wilford 1966), but is less acutely toxic to warm-water fish species (*Lepomis macrochirus* and *Ictalurus punctatus*). The toxicity to antimalarial drugs has been studied primarily in fish; results indicate that no compound is significantly more toxic than any other. In limited studies, invertebrates appear to be the most sensitive trophic group to antimalarial drugs (Johnson 1976; Calleja et al. 1994; Lilius et al. 1994). Metronidazole is the only antiprotozoal compound studied thus far, and this drug has only demonstrated acute toxicity to algae (12.5–39.1 mg/L) (Lanzky and Halling-Sorenson 1997); however, it was not acutely toxic to invertebrates or fish (Wilford 1966). The antiviral or retroviral compounds didanosine, famciclovir, stavudine, and zalcitabine have been studied in *Daphnia* spp. and are not acutely toxic at levels below 820 mg/L (FDA-CDER 1996). Excluding algae, all antibiotic compounds are relatively nontoxic to aquatic invertebrates and fish. Unfortunately, very little research has been thus far conducted on the effects of other antibiotic compounds. Although algae are sensitive in short-term exposures, it is unlikely that effects will be observed at the levels antibiotics are realistically expected to occur in the environment.

2.6 Antidepressants

Eight different antidepressants have been tested for aquatic effects, and sertraline was identified as being the most acutely toxic antidepressant across all trophic groups (i.e., algae, invertebrates, fish, and amphibians). Sertraline is in the SSRI class of antidepressants, and is acutely lethal (LC_{50}) to algae at concentrations >0.01 mg/L, to invertebrates above 0.1 mg/L and to fish at concentrations greater than 0.3 mg/L (Henry et al. 2004; Richards and Cole 2006; Christensen et al. 2007; Heckmann et al. 2007; Valenti et al. 2009). Generally, SSRI antidepressants are more toxic than tricyclic lithium and selective norepinephrine reuptake inhibiting-based antidepressants. Similar to antibiotics, algae appears to be the trophic groups that are most sensitive to SSRI and tricyclic antidepressants; acute toxicity values are observed in the high µg/L and low mg/L range, although actual environmental exposure concentrations are too low to produce acute toxicity (Monteiro and Boxall 2010). In addition to death, short-term exposure to SSRIs can produce sub-lethal effects. One such effect was observed in a unionid mussel, in which a 48-h exposure to fluoxetine produced a significantly increased nonviable glochidia released by *Ellipto complanata* (Bringolf et al. 2010). Although effects have been observed in the laboratory as a result of short-term antidepressant exposure, acute effects at realistic exposures levels are not expected.

2.7 Antidiabetics

Acarbose, an α-glucosidase inhibitor (Walton et al. 1979), and metformin, which suppresses hepatic gluconeogenesis (Kirpichnikov et al. 2002), have been tested for acute toxicity in invertebrates, fish, and algae (FDA-CDER 1996). Acarbose was not acutely toxic to either invertebrates or fish at the highest concentration studied (1,000 mg/L; FDA-CDER 1996). Metformin was acutely toxic to *D. magna,* displaying a 48-h EC_{50} (immobility) of 64 mg/L (Cleuvers 2003). However, *L. macrochirus* and *D. subspicatus* were not acutely sensitive to metformin toxicity at the highest concentration studied (982 mg/L and 320 mg/L, respectively; Cleuvers 2003) thus, no aquatic risk is likely.

2.8 Antiepilectics

The acute toxicity of carbamazepine has been investigated for bacteria, invertebrates, fish, amphibians, algae, and natural communities of biofilms. Biofilms, which are aggregates of microorganisms, appear to be the most sensitive aquatic test system to carbamazepine. Exposure to carbamazepine caused a decreased volume of biofilm at concentrations twofold lower than concentrations needed to induce adverse effects in any other species (Lawrence et al. 2005). The bacterium *V. fischeri*, in the Microtox® assay, does not appear acutely sensitive to carbamazepine, and displays 5-min EC_{50} (illuminescence) values between 52.2 and 81 mg/L (Ferrari et al. 2004; Kim et al. 2007); this bacterium also does not appear to be sensitive to longer periods of exposure (Ferrari et al. 2003). *Ceriodaphnia dubia* are more sensitive than *D. magna,* when comparing immobility after 48-h (EC_{50}); however, neither species would be considered to be sensitive (Cleuvers 2003; Ferrari et al. 2003; Kim et al. 2007). *Oryzias latipes* is the only fish species studied and was more sensitive to carbamazepine than were most other species studied. The growth and density of biofilms were the most sensitive endpoint studied for carbamezapine (Kim et al. 2007). Carbamazepine had no effect on embryonic development in the commonly used amphibian model species, *Xenopus laevis*, at the highest concentration investigated (100 mg/L; Richards and Cole 2006). Gabapentin, an antiepileptic adjunctive and neuropathic pain reliever, was not acutely toxic up to 1,100 mg/L in the only organism tested (*D. magna*; FDA-CDER 1996).

2.9 Antihypertensives

Thirteen different antihypertensive pharmaceuticals have been investigated for acute toxicity; however, the acute toxicity of only a few compounds has been examined in more than a single species. For invertebrates, the β-adrenergic receptor blocker propranolol and the enantiomer *R*-propranolol are reported to be the most toxic antihypertensive drugs, having EC/LC$_{50}$ values ranging between 1.38 and 17.7 mg/L; however, these concentrations are greater than observed environmental concentrations (Calleja et al. 1994; Lilius et al. 1994; Huggett et al. 2002; Cleuvers 2003; Stanley et al. 2006). Huggett et al. (2002) identified that β-blocker acute toxicity to the cladocerans *D. magna* and *C. dubia* was related to increasing log *P* values, suggesting that the acute toxicity of beta blockers to invertebrates is elicited through narcosis. Because propranolol is distributed as a racemic mixture, Stanley et al. (2006) compared the acute toxicity of the racemic mixture (*R/S*-propranolol) to each enantiomer (*R*-propranolol, *S*-propranolol), using *D. magna* and *P. promelas* as a model invertebrate and vertebrate, respectively. Their results indicate that no appreciable toxicity differences among chemical treatments or organisms existed between enantiomers. The acute toxicity of diltiazem, a calcium channel blocker, was investigated in bacteria (*V. fischeri*), invertebrates (*D. magna*), and fish (*O. latipes*) (Kim et al. 2007). Fish were reported to be generally less acutely sensitive than invertebrates and algae to calcium channel and beta blockers (Huggett et al. 2002; Stanley et al. 2006; Kim et al. 2007).

2.10 Antineoplastics

Six antineoplastic compounds have been investigated for acute toxicity. Tamoxifen appears to be most toxic, showing LC$_{50}$ values almost 100× lower than other antineoplastic compounds, although only invertebrates were investigated (Kang et al. 2002). Methotrexate has been studied in the greatest number of species of algae, invertebrate, and fish. Although *S. subspicatus* was most sensitive, toxicity values exceeded 250 mg/L for all tested species (Henschel et al. 1997). The remaining antineoplastic compounds (cladribine, paclitaxel, and thiotepa) were not toxic to *Daphnia* spp. and had EC_{50} and LC_{50} values greater than the highest tested concentration (250 mg/L; FDA-CDER 1996; Henschel et al. 1997).

2.11   Antipsychotics

Thioridazine is a phenothiazine antipsychotic drug used for the treatment of psychotic disorders. Thioridazine's acute toxicity has been studied in *B. calyciflorus*, *D. magna*, and *S. proboscideus*. Thioridazine was acutely toxic to all three species, exhibiting 24-h EC_{50} and LC_{50} values <1 mg/L, but these concentrations were still greater than environmental concentrations (Calleja et al. 1994; Lilius et al. 1994). Risperidone, another antipsychotic drug used to treat schizophrenia, was acutely toxic to *D. magna* (EC_{50}) and to *L. macrochirus* (LC_{50}) at 6.0 mg/L (FDA-CDER 1996).

2.12   Antiulceratives

Five different antiulcerative pharmaceuticals were investigated for acute toxicity; however, only cimetidine has been extensively studied. Cimetidine is a histamine H_2-receptor antagonist that inhibits production of stomach acid. Cimetidine is the only antacid that has been investigated for its toxic effects to aquatic organisms, and only short-term acute toxicity tests have been conducted. Toxicity testing thus far indicates cimetidine has very little acute toxicity to bacteria (*V. fischeri*), invertebrates (*D. magna*), and fish (*O. latipes*) (Kim et al. 2007). The acute toxicity of other antiulcerative compounds, such as omeprazole, a proton pump inhibitor, have been studied in *Dapnia* spp., and results demonstrate that omeprazole is the most acutely toxic proton pump inhibitor to invertebrates (FDA-CDER 1996). Conversely, lansoprazole, also a proton pump inhibitor, was significantly more toxic to *O. mykiss*, than to any other antiulcerative pharmaceutical studied thus far (FDA-CDER 1996). There are currently no studies published in which the chronic toxicity of cimetidine, or other antacids and antiulcerative compounds have been studied, and, although acute risk is unlikely, the risk cannot yet conclusively be evaluated.

2.13   Bone Resorption Inhibitors

Alendronate sodium, etidronic acid, and tiludronate disodium are used as treatments for osteoporosis and other bone disorders. Alendronate sodium and etidronic acid have been investigated for acute toxicity in invertebrates, fish, and algae although based on acute toxicity no risk is expected. Etidronic acid is more acutely toxic to fish than is alendronate sodium; however, it appears the opposite is true for invertebrates (Gledhill and Feijtel 1992; FDA-CDER 1996). Algae were not acutely sensitive to either alendronate sodium or etidronic acid. Acute effects of tiludronate disodium has only been investigated for *D. magna,* for which a 48-h LC_{50} value of 320 mg/L (Sanofi 1996) was reported; this value was similar to values observed for etidronic acid. Similar to many other pharmaceutical compounds, bone resorption inhibitors are most acutely toxic to algae, although no known receptor exists in phytoplankton.

2.14   Cholinergic Agonists

The cholinergic agonists, nicotine sulfate and nicotine, are relatively toxic to invertebrates and fish over short exposure durations. *D. magna* is the only invertebrate thus far tested and observed to be sensitive to nicotine (i.e., 24-h EC_{50} (immobilization) of 3.0 mg/L). Fish (*L. machrochirus* and *P. promelas*) have also been investigated for nicotine sensitivity and were found to have LC_{50} values ranging between 4 and 20 mg/L, which is significantly greater than observed environmental concentrations (FDA-CDER 1996). Nicotine sulfate and nicotine were equally toxic to *P. promelas* (Russom et al. 1997).

2.15   Diuretics

To date the acute toxicity of the diuretics has been investigated in only one study. In this study, bendroflumethiazide and furosemide were evaluated for effects in the cnidarian *Hydra vulgaris.* The authors of this study did not identify any acute toxicity during a 7-day exposure at the highest concentration studied (100 mg/L; Pascoe et al. 2003).

2.16   Estrogens and Antiestrogens

The acute toxicity of female hormones has been tested in a variety of organisms and trophic groups; however, estrogens are the only pharmaceutical category in which more chronic than acute studies have been performed. The prototypical estrogen compound diethylstilbestrol (DES) has been well studied, in that four studies have been conducted on six different species. Coats et al. (1976) investigated the acute toxicity of DES on algae, invertebrates, and fish, identifying invertebrates (*Culex pipiens* and *D. magna*) as the most sensitive group (48-h $LC_{50} = 4$ mg/L). Baldwin et al. (1995) and Zou and Fingerman (1997) reported *D. magna* to be even more sensitive than previously reported to DES exposure, based on 48-h LC_{50} values ranging from 1.1 to 1.2 mg/L; however, EE2 is significantly more acutely toxic than DES to algae ($EC_{50}=0.84$ mg/L) and fish (96-h $EC_{50}=1.6$ mg/L) (Kopf 1995; Schweinfurth et al. 1996). For *D. magna*, Clubbs and Brooks (2007) reported a 96-h EC_{50} mortality value of 2.7 mg/L for EE2, but a much lower value of 127 µg/L for the selective estrogen receptor modulator faslodex, highlighting the relatively higher acute toxicity of antiestrogens compared to synthetic estrogens. Based on observed environmental concentrations, it is unlikely that acute risk would result from the typical exposure of freshwater organisms to synthetic estrogens.

2.17   Lipid-Lowering Drugs

Many lipid-lowering/antihyperlipoproteinemic drugs have been investigated for potential acute toxicity. The effects of lipid-lowering drugs have been investigated in all trophic groups except plants. No trophic group appears more sensitive than any other group. Clofibrate appears more toxic than any other lipid-lowering drug, including clofibrinic acid, which is the active metabolite of clofibrate (Kopf 1995; Henschel et al. 1997; Cleuvers 2003). The additional lipid-lowering drugs bezafibrate, fenofibrate, and gemfibrozil have been investigated for toxicity in invertebrates and algae, but all of these were relatively nontoxic and would not be expected to cause acute toxicity at environmentally relevant concentrations.

2.18   Stimulants

Two central nervous system stimulants, amphetamine and caffeine, have been investigated for acute toxicity. Acute testing was carried out in four invertebrates and a single fish species. In all studies, amphetamine was between 2 and 100× more toxic to invertebrates and 6× more toxic to fish (*P. promelas*), in comparison to caffeine (Calleja et al. 1994; Lilius et al. 1994; Russom et al. 1997). For both compounds, *D. magna* was the most sensitive species, and had 24-h EC_{50} (immobility) values of 60 and 160 mg/L for amphetamine and caffeine, respectively. Caffeine has been observed to cause a stimulatory effect in biofilm communities, and in particular, low levels (10 µg/L) of caffeine exposure caused increased growth and cell volume of biofilms (Lawrence et al. 2005), although these effects occurred at levels greater than observed environmental concentrations.

2.19   Sedatives

Four barbituates: amobarbital, pentobarbital, phenobarbital, and secobarbital have been investigated for acute toxicity. Based on limited data, secobarbatal is the most toxic sedative to fish (Russom et al. 1997). Invertebrate sensitivity has only been tested in one study, the results of which indicated invertebrates were not sensitive to phenobarbital (24-h EC and LC_{50} values between 1,212 and 5,179 mg/L; Calleja et al. 1994). Several other sedatives with varying modes of action were tested for acute toxicity in freshwater aquatic organisms. The most toxic of these compounds was the benzodiazepine midazolam, which had an EC_{50} value of 0.2 mg/L in *D. magna* (FDA-CDER 1996). Currently, diazepam and ondansetron are the most studied sedatives. Diazepam's acute toxicity has been elucidated in several aquatic invertebrate species, with *D. magna* being the most sensitive specie examined thus far (Calleja et al. 1994; Lilius et al. 1994; Pascoe et al. 2003). Similarly, orphenadrine effects have only been examined in invertebrates, and the most sensitive species was the fairy shrimp *S. proboscideus* (Calleja et al. 1994). Although not all trophic groups have been examined for sedative effects, acute risk is not expected based on previous results.

2.20 Other Compounds

Atropine sulfate is an anticholinergic pharmaceutical that has been studied for acute toxicity in three different freshwater invertebrates. *S. proboscideus*, *D. magna*, and *B. calyciflorus* are not sensitive to atropine sulfate; their EC and LC_{50} values were >300 mg/L in one study (Calleja et al. 1994) and >200 mg/L in *D. magna* in another (Lilius et al. 1994).

Dorzolamide is a carbonic anhydrase inhibitor used for the treatment of glaucoma. The acute toxicity testing of dorzolamide in *D. magna* and *P. promelas* indicates that these species are not sensitive to dorzolamide exposure over short-term exposures (i.e., EC_{50} and LC_{50} values for *D. magna* and *P. promelas* of 699 and >1,000 mg/L, respectively) (FDA-CDER 1996).

Iopromide is used as nonionic contrast media, and has been tested for acute toxicity in several algae, invertebrate, and fish species. Thus far, no acute toxicity has been observed at the highest levels (>137 mg/L) investigated (FDA-CDER 1996; Schweinfurth et al. 1996).

The anticoagulant warfarin has been examined for acute toxicity in several aquatic invertebrates and fish. Warfarin was originally marketed as a rat poison, but is now used to treat blood-clotting disorders (i.e., Coumadin®). For freshwater aquatic invertebrates, 24-h LC_{50} values ranged from 342 to 444 mg/L for *S. proboscideus* and *B. calciflorus* (Calleja et al. 1994), respectively, and caused immobilization in *D. magna* between 89 and 475 mg/L (Calleja et al. 1994; Lilius et al. 1994). The fish species *Rasbora heteromorpha* is significantly more sensitive than invertebrates, which probably results from the warfarin's effects on vitamin K epoxide reductase, which is found primarily in vertebrates (Tooby et al. 1975).

Diphenhydramine, an H1-blocking antihistamine, has been detected in multiple environmental matrices, including aquatic organisms (Ramirez et al. 2007, 2009). In addition to antihistamine activity, the 5-HT (5-hydroxytryptamine) reuptake blocking activity of diphenhydramine provided the impetus for development of the SSRI antidepressants (Wong et al. 2005). Further, diphenhydramine also inhibits the acetylcholine receptor, and thus has been used to treat organophosphorus pesticide poisoning (Bird et al. 2002). The acute toxicity (48-h) of diphenhydramine was markedly greater to *D. magna* ($EC_{50} = 0.374$ mg/L) than to *P. promelas* ($LC_{50} = 2.09$ mg/L) (Berninger et al. 2011), ergo further studies of antihistamines are needed.

3   Chronic Toxicity of Pharmaceuticals in Aquatic Organisms

The chronic toxicity of pharmaceuticals to aquatic species has been studied for at least 65 individual compounds that comprise more than 20 pharmaceutical classes (Table 2). Data have been compiled for most trophic groups, including invertebrates, fish, amphibians, plants, and algae. Currently, most chronic toxicity data are available for invertebrates and fish, whereas only limited data exists for amphibians and benthic invertebrates (infaunal or epifaunal), including bivalves. Additionally, available chronic data rarely reflected investigations that addressed the key targets in different organisms or different life stages; most chronic data are derived from standardized chronic toxicity testing methods or protocols.

3.1   Analgesics

Five analgesic compounds, including one non-narcotic drug and four NSAIDs (acetaminophen, acetylsalicylic acid, diclofenac, ibuprofen, and naproxen) and three breakdown products (gentisic acid, O-hydroxyhippuric acid, and salicylic acid), have been investigated to identify potential long-term toxic effects in aquatic organisms. The mechanism of action for vertebrates is through the inhibition of cyclooxygenase (COX-1 or COX-2), which is effected by inhibiting prostaglandin synthesis (Vane and Botting 1998). This mechanism indicates that fish and amphibians would be the most susceptible trophic groups, because they have cyclooxygenase enzymes. Although acetaminophen is not currently classified as an NSAID, it has a similar mechanism of action in inhibiting prostaglandin synthesis, likely through the COX-2 enzyme (Hinz et al. 2008). The chronic toxicity of analgesic compounds has been tested in invertebrates, fish, amphibians, fungi, and plants. Specifically, 12 chronic studies have been performed on ibuprofen, which has been the most commonly studied analgesic. These studies encompassed 11 species; however, acetaminophen was the most thoroughly studied individual compound, because both parent and metabolite products were examined. Chronic toxicity studies, using several invertebrates (viz., *D. magna*, *Daphnia longispina*, *Ceriodaphnia dubia*, *Brachionus calyciflorus*, *Planorbis carinatus*, and *H. vulgaris*), focused entirely on the parameters of survival, growth, and reproduction.

Human Pharmaceuticals in the Aquatic Environment… 43

Table 2 Chronic toxicity data for human-use pharmaceuticals

Compound	Category	Species	Trophic group	Endpoint/Duration	LOEC (µg/L)	NOEC (µg/L)	Reference
Acetylsalicylic acid (aspirin)	Analgesic	H. vulgaris	Invert.	Survival; Feeding, Bud formation, Regeneration	No effect (17-day)		Pascoe et al. (2003)
		D. longispina	Invert.	21-day reproduction	5,600		Marques et al. (2004)
		D. magna	Invert.	21-day reproduction	>10,000		Marques et al. (2004)
Diclofenac	Analgesic	V. fisheri	Bacteria	30-min EC_{50}	11,454		Ferrari et al. (2003)
		B. calyciflorus	Invert.	48-h Reproduction	125,000	25,000	Ferrari et al. (2003)
		C. dubia	Invert.	7-day Reproduction	2,000	1,000	Ferrari et al. (2003)
		O. mykiss	Fish	28-day Liver/gill histology	1		Triebskorn et al. (2004)
		O. mykiss	Fish	28-day Digestive tract histopathology	5 (kidney/gills)		Schwaiger et al. (2004)
		O. latipes	Fish	9-day Feeding Behavior	1,000		Nassef et al. (2010)
		D. rerio	Fish	10-day Repro/Embryo Survival	8,000	4,000	Ferrari et al. (2003)
		S. trutta	Fish	21-day Histopathology		0.5	Hoeger et al. (2005)
		L. minor	Plant	7-day growth	7,500 (EC_{50})		Cleuvers (2003)
Gentisic acid (aspirin metabolite)	Analgesic	D. longispina	Invert.	21-day reproduction	320		Marques et al. (2004)
		D. magna	Invert.	21-day reproduction	320		Marques et al. (2004)
Ibuprofen	Analgesic	S. aureus	Bacteria	Growth		150,000	Elvers and Wright (1995)
		D. magna	Invert.	14-day development, 14-day reproduction, 14-day survival	40,000 (Develop.); 20,000 (Repro.);80,000 (Survival)	20,000 (Population growth)	Heckmann et al. (2007)
		G. pulex	Invert.	Behavior	0.01		De Lange et al. (2006)

(continued)

Table 2 (continued)

Compound	Category	Species	Trophic group	Endpoint/Duration	LOEC (μg/L)	NOEC (μg/L)	Reference
		H. vulgaris	Invert.	Survival; feeding, bud formation, regeneration	No effect (17-day)		Pascoe et al. (2003)
		P. carinatus	Invert.	21-day survival, growth, reproduction	24,300 (Growth)	53,600, 10,200, 24,300	Pounds et al. (2008)
		O. latipes	Fish	6-week reproduction	1.00 (TEC) [a]		Flippin et al. (2007)
		O. mykiss	Fish	8-day ion regulation	1000 (TEC)		Gravel et al. (2009)
		D. rerio	Fish	6-day cardiac abnormalities	>10 (TEC)		David and Pancharata (2009)
		D. fungi	Fungi	MIC, growth		200,000	Sanyal et al. (1993)
		L. gibba	Plant	7-day growth	>1,000 (EC$_{10}$)		Brain et al. (2004a)
		L. minor	Plant	7-day growth	22,000 (EC$_{50}$)		Cleuvers (2003)
		L. minor	Plant	7-day growth	10,000 (LOEC)		Pomati et al. (2004)
Naproxen	Analgesic	C. dubia	Invert.	7-day population growth (EC$_{50}$)	330		Isidori et al. (2005a)
		B. calyciflorus	Invert.	48-h growth (EC$_{50}$)	560		Isidori et al. (2005a)
		H. attenuata	Invert.	96-h morphology	5,000	1,000	Quinn et al. (2008)
		L. minor	Plant	7-day growth	24,200 (EC$_{50}$)		Cleuvers (2003)
Paracetamol/ Acetaminophen	Analgesic	V. fischeri	Bacteria	30-min EC$_{50}$	650,000		Henschel et al. (1997)
		M. galloprovincialis	Invert.	10-day behavior feeding rate	>403	403	Sole et al. (2010)
		R. pipiens	Amph.	28-day survival, growth, behavior	No effect >1,000		Fraker and Smith (2004)
		B. americanus	Amph.	14-day survival, growth, behavior	100 (survival) No effect (growth) 1 (behavior)		Smith and Burgett (2005)
		L. gibba	Plant	7-day growth	>1,000 (EC$_{10}$)		Brain et al. (2004b)

Compound	Class	Species	Type	Endpoint	Value		Reference
O-Hydroxyhippuric acid (aspirin metabolite)	Analgesic	D. longispina	Invert.	21-day reproduction	84,500		Marques et al. (2004)
		D. magna	Invert.	21-day reproduction	186,000		Marques et al. (2004)
Salicylic acid (aspirin metabolite)	Analgesic	V. fischeri	Bacteria	30-min EC$_{50}$	90,000		Henschel et al. (1997)
Methyl testosterone	Androgen	D. longispina	Invert.	21-day reproduction	5,600		Marques et al. (2004)
		D. magna	Invert.	21-day reproduction	>10,000–<20,000		Marques et al. (2004)
		L. stagnalis	Invert.	12-week reproduction	1		Czech et al. (2001)
		M. cornuarietis	Invert.	12-month sex development	<0.1		Schulte-Oehlmann et al. (2004)
		D. magna	Invert.	21-day sex ratio	8 µM		Mu and LeBlanc (2002)
		C. carassius	Fish	80-day phenotypic sex ratio	0.01		Fujioka (2002)
		P. promelas	Fish	13-day phenotypic sex	10		Zerulla et al. (2002)
		O. latipes	Fish	42-day sex reversal	<0.01		Pascoe et al. (2003)
Amoxicillin	Antibiotic	H. vulgaris	Invert.	Survival; feeding, bud formation, regeneration	No effect (17-day)		
Cephalexin	Antibiotic	L. gibba	Plant	7-day Wet Mass	>1,000 (EC$_{10}$)		Brain et al. (2004a)
Chlortetracycline	Antibiotic	L. gibba	Plant	7-day growth	36 (EC$_{10}$), 219 (EC$_{50}$)		Brain et al. (2004a)
Ciprofloxacin	Antibiotic	L. gibba	Plant	7-day growth	106 (EC$_{10}$), 698 (EC$_{50}$)		Brain et al. (2004a)
Clarithromycin	Antibiotic	L. minor	Plant	7-day Growth	203 (EC$_{50}$)		Robinson et al. (2005)
		D. magna	Invert.	21-day reproduction	40 (EC$_{50}$)		Robinson et al. (2005)
		D. magna	Invert.	21-day reproduction	6.3		Robinson et al. (2005)
		B. calyciflorus	Invert.	48-h population growth	12,210 (EC$_{50}$)	3.1	Isidori et al. (2005b)
		C. dubia	Invert.	7-day Population Growth	8,160 (EC$_{50}$)		Isidori et al. (2005b)
Clinafloxacin	Antibiotic	L. minor	Plant	7-day growth rate	62 (EC$_{50}$)		Robinson et al. (2005)

(continued)

Table 2 (continued)

Compound	Category	Species	Trophic group	Endpoint/Duration	LOEC (µg/L)	NOEC (µg/L)	Reference
Doxycycline	Antibiotic	L. gibba	Plant	7-day growth	55 (EC_{10}), 316 (EC_{50})		Brain et al. (2004a)
Enrofloxacin	Antibiotic	D. magna	Invert.	21-day survival	11,470 (EC_{50})		Park and Choi (2008)
		D. magna	Invert.	21-day reproduction	15,000	5,000	Park and Choi (2008)
		L. minor	Plant	7-day growth rate	114 (EC_{50})		Robinson et al. (2005)
Erythromycin	Antibiotic	C. dubia	Invert.	7-day population growth	220 (EC_{50})		Isidori et al. (2005b)
		B. calyciflorus	Invert.	48-h population growth	940 (EC_{50})		Isidori et al. (2005b)
		D. magna	Invert.	21-day survival		248	Meinertz et al. (2010)
		L. minor	Plant	7-day growth	100,000		Pomati et al. (2004)
		L. minor	Plant	7-day Growth	5,620 (EC_{50})		Pomati et al. (2004)
		L. gibba	Plant	7-day wet mass	121 (EC_{10})		Brain et al. (2004a)
Flumequine	Antibiotic	L. minor	Plant	7-day growth rate	2,470 (EC_{50})		Robinson et al. (2005)
Levofloxacin	Antibiotic	D. magna	Invert.	21-day reproduction	340 (EC_{50})		Robinson et al. (2005)
		D. magna	Invert.	21-day reproduction	63	31	Robinson et al. (2005)
		L. gibba	Plant	7-day wet mass	13 (EC_{10}), 185 (EC_{50})		Brain et al. (2004a)
		L. minor	Plant	7-day wet mass	51 (EC_{50})		Robinson et al. (2005)
Lincomycin	Antibiotic	B. calyciflorus	Invert.	48-h population growth	680 (EC_{50})		Isidori et al. (2005b)
		C. dubia	Invert.	7-day population growth	7,200 (EC_{50})		Isidori et al. (2005b)
		L. gibba	Plant	7-day growth	>1,000 (EC_{10})		Brain et al. (2004a)
Lomefloxacin	Antibiotic	L. gibba	Plant	7-day wet mass	8 (EC_{10}), 97 (EC_{50})		Brain et al. (2004a)
		L. minor	Plant	7-day wet mass		106 (EC_{50})	Robinson et al. (2005)
Metronidazole	Antibiotic	D. magna	Invert.	21-day reproduction		250,000	Wollenberger et al. (2000)
Neomycin	Antibiotic	L. gibba	Plant	7-day wet mass	>1,000 (EC_{50})		Brain et al. (2004a)

Compound	Type	Species	Group	Endpoint	Value		Reference
Norfloxacin	Antibiotic	*L. gibba*	Plant	7-day wet mass	206 (EC$_{10}$), 913 (EC$_{50}$)		Brain et al. (2004a)
Ofloxacin	Antibiotic	*C. dubia*	Invert.	7-day reproduction	10,000 (LOEC)		Ferrari et al. (2003)
		C. dubia	Invert.	7-day population growth	3,130 (EC$_{50}$)		Isidori et al. (2005b)
		B. calyciflorus	Invert.	48-h reproduction	12,500		Ferrari et al. (2003)
		B. calyciflorus	Invert.	48-h population growth	530 (EC$_{50}$)		Isidori et al. (2005b)
		L. gibba	Plant	7-day growth	121 (EC$_{10}$), 532 (EC$_{50}$)		Brain et al. (2004a)
		L. minor	Plant	7-day wet mass	126 (EC$_{50}$)		Robinson et al. (2005)
Oxolinic acid	Antibiotis	*D. magna*	Invert	21-day reproduction		380	Wollenberger et al. (2000)
Oxytetracycline	Antibiotic	*V. fischeri*	Bacteria	30-min bioluminescence	64,500 (EC$_{50}$)		Isidori et al. (2005b)
		D. magna	Invert.	21-day reproduction	46,200 (EC$_{50}$)		Wollenberger et al. (2000)
		C. dubia	Invert.	7-day population growth	180 (EC$_{50}$)		Wollenberger et al. (2000)
		B. calyciflorus	Invert.	48-h population growth	1,870 (EC$_{50}$)		Wollenberger et al. (2000)
		L. minor	Plant	7-day growth	4,920 (EC$_{50}$)		Pro et al. (2003)
		L. gibba	Plant	7-day growth	788 (EC$_{10}$), 1,010 (EC$_{50}$)		Brain et al. (2004a)
Roxithromycin	Antibiotic	*L. gibba*	Plant	7-day wet mass	>1,000 (EC$_{10}$)		Brain et al. (2004a)
Streptomycin	Antibiotic	*L. gibba*	Plant	7-day wet mass	>1,000 (EC$_{10}$)		Brain et al. (2004a)
Sulfachlor-pyridazine	Anibiotic	*L. minor*	Plant	7-day growth	2,330 (EC$_{50}$)		Pro et al. (2003)
Sulfadiazine	Antibiotic	*D. magna*	Invert.	21-day reproduction	13,700 (EC$_{50}$)		Robinson et al. (2005)
Sulfadi-methazine	Antibiotic	*D. magna*	Invert.	21-day reproduction	3,125	1,563	De Liguoro et al. (2009)
Sulfadi-methoxine	Antibiotic	*L. gibba*	Plant	7-day wet mass	44 (EC$_{10}$), 248 (EC$_{50}$)		Brain et al. (2004a)
Sulfamethazine	Antibiotic	*L. gibba*	Plant	7-day wet mass	>1,000 (EC$_{10}$)		Brain et al. (2004a)

(continued)

Table 2 (continued)

Compound	Category	Species	Trophic group	Endpoint/Duration	LOEC (µg/L)	NOEC (µg/L)	Reference
Sulfamethoxazole	Antibiotic	V. fischeri	Bacteria	30-min bioluminescence	23,300 (EC_{50})		Isidori et al. (2005b)
		B. calyciflorus	Invert.	48-h reproduction	25,000		Ferrari et al. (2003)
		B. calyciflorus	Invert.	48-h population growth	9,630 (EC_{50})		Isidori et al. (2005b)
		C. dubia	Invert.	7-day reproduction	250		Ferrari et al. (2003)
		C. dubia	Invert.	7-day population growth	210 (EC_{50})		Isidori et al. (2005b)
		L. gibba	Plant	7-day wet mass	81 (EC_{50})	11 (EC_{10})	Brain et al. (2004a)
		L. gibba	Plant	7-day wet mass, 7-day frond number	61.6 (EC_{50}), 132 (EC_{50})	21 (EC_{10}), 58 (EC_{10})	Brain et al. (2008a, b)
Sulfathiazole	Antibiotic	D. magna	Invert.	21-day reproduction	35,000	11,000	Park and Choi (2008)
Tetracycline	Antibiotic	D. magna	Invert.	21-day reproduction	44,800 (EC_{50})		Wollenberger et al. (2000)
		L. minor	Plant	7-day growth	1,000 (LOEC)		Elvers and Wright (1995)
		L. minor	Plant	7-day growth	1,060 (EC_{50})		Pomati et al. (2004)
		L. gibba	Plant	7-day wet mass	230 (EC_{10}), 723 (EC_{50})		Brain et al. (2004a)
Trimethoprim	Antibiotic	D. magna	Invert.	21-day reproduction	20,000	6,000	Park and Choi (2008)
		L. gibba	Plant	7-day growth	>1,000 (EC_{10})		Smith and Burgett (2005)
Tylosin	Antibiotic	D. magna	Invert.	21-day reproduction	45,000 (EC_{50})		Wollenberger et al. (2000)
		L. gibba	Plant	7-day wet mass	>1,000 (EC_{10})		Brain et al. (2004a)
Citalopram	Antidepressant	C. dubia	Invert.	7-day reproduction	800 (LOEC)		Henry et al. (2004)
Fluoxetine	Antidepressant	C. dubia	Invert.	7-day reproduction		89	Henry et al. (2004)
		C. dubia	Invert.	7-day reproduction	112	56	Brooks et al. (2003)
		D. magna	Invert.	21-day reproduction	178	89	Brooks et al. (2005)
		D. magna	Invert.	21-day offspring length	31	8.9	Pery et al. (2008)

Species	Type	Endpoint	Value		Reference
D. magna	Invert.	30-day reproduction	36 (Stimulation)		Flaherty et al. (2001)
H. azteca	Invert.	7-day reproduction	>43,000		Brooks et al. (2003)
E. marinus with acanthocephalan infection	Invert.	21-day behavior	0.1		Guler et al. (2010)
P. antipodarum	Invert.	Reproduction	69	13	Pery et al. (2008)
P. antipodarum	Invert.	56-day embryo	0.81 (EC$_{10}$)	0.47	Nentwing (2007)
G. pulex	Invert.	Behavior	0.1		De Lange et al. (2006)
C. tentans	Invert.	10-day growth	1,300		Brooks et al. (2003a)
C. tentans	Invert.	10-day survival	15,200 (LC$_{50}$) 1,300 (LOEC)		Brooks et al. (2003b)
H. azteca	Invert.	Growth	5.4		Brooks et al. (2003a)
H. azteca	Invert.	28-day growth	100	33	Pery et al. (2008)
H. azteca	Invert.	10-day growth	5,600		Brooks et al. (2003b)
O. latipes	Fish	28-day estradiol conc.	0.1–0.5		Huggett et al. 2003, Foran et al. (2004)
G. affinis	Fish	Survival, behavior; survival, sex ratio, development	5 development		Henry and Black (2008)
G. affinis	Fish	7-day survival	546 (LC$_{50}$)		Henry and Black (2008)
P. promelas	Fish	7-day behavior (feeding), growth	51, 51		Stanley et al. (2007)
C. auratus	Fish	28-day feeding rate	54		Mennigen et al. (2010)
C. auratus	Fish	14-day male reproductive physiology	54		Mennigen et al. (2010b)
X. laevis	Amph.	Metamorphosis; Growth	>10; 10		Conners et al. (2009)
R. pipiens	Amph.	50-day development	0.029		Foster et al. (2010)
L. minor	Plant	7-day Growth	>1,000 (EC$_{10}$)		Brain et al. (2004a)
L. gibba	Plant	7-day GROWTH	>1,000 (EC$_{10}$)		Brain et al. (2004b)

(continued)

Table 2 (continued)

Compound	Category	Species	Trophic group	Endpoint/Duration	LOEC (µg/L)	NOEC (µg/L)	Reference
Fluvoxamine	Antidepressant	V. fischeri	Bacteria	30-min Bioluminescence	10,720 (EC_{50})		Minagh et al. (2009)
		V. fischeri	Bacteria	30-min bioluminescence	4,500	2,250	Minagh et al. (2009)
		D. magna	Invert.	21-day reproduction	100	32	Minagh et al. (2009)
		D. magna	Invert.	21-day survival	100	32	Minagh et al. (2009)
		C. dubia	Invert.	7-day reproduction	366		Henry et al. (2004)
Paroxetine	Antidepressant	C. dubia	Invert.	7-day reproduction	220		Henry et al. (2004)
Sertraline	Antidepressant	D. magna	Invert.	21-day lethality 21-day reproduction	100 (LOEC)	32 (NOEC)	Heckmann et al. (2007)
		C. dubia	Invert.	7-day reproduction	45 (LOEC)		Henry et al. (2004)
		L. gibba	Plant	7-day growth	>1,000 (EC_{50})		Brain et al. (2004b)
Metformin	Antidiabetic	L. minor	Plant	7-day growth	110,000 (EC_{50})		Cleuvers (2003)
Carbamazepine	Antiepilectic	V. fischeri	Bacteria	30-min bioluminescence	>81,000 (EC_{50})		Ferrari et al. (2003)
		B. calyciflorus	Invert.	48-h reproduction	754	377	Ferrari et al. (2003)
		C. riparius	Invert.	28-day emergence	0.625 (mg/kg)		Nentwig et al. (2004)
		L. variegatus	Invert.	28-day reproduction	>10 (mg/kg)		Nentwig et al. (2004)
		C. dubia	Invert.	7-day reproduction	100	25	Ferrari et al. (2003)
		C. tentans	Invert.	10-day survival, growth	47,300, 9,500 (LC_{50}, EC_{50})	9,500, 2,600 (LC_{10}, EC_{10})	Dussault et al. (2008)
		H. azteca	Invert.	10-day survival, growth	9,900, 600 (LC_{50}, EC_{50})	1,500, 2,400 (LC_{10}, EC_{10})	Dussault et al. (2008)
		G. pulex	Invert.	Behavior	0.01		De Lange et al. (2006)
		D. rerio	Fish	10-day hatching/larvae mor.	50,000	25,000	Ferrari et al. (2003)

Compound	Class	Species	Group	Endpoint	Value	Value	Reference
ZM 189, 154	Antiestrogen	O. mykiss	Fish	21-day pathology	>100 (Liver), 1 (Kidney)		Triebskorn et al. (2007)
		O. latipes	Fish	9-day behavior	6,150		Nassef et al. (2010)
		L. gibba	Plant	7-day growth	>1,000 (EC_{10})		Brain et al. (2004b)
		L. minor	Plant	7-day growth	25,500 (EC_{50})		Cleuvers (2003)
		D. rerio	Fish	38-day VTG, 60-day sex ratio	100 (LOEC)		Andersen et al. (2004)
Faslodex	Antiestrogen	D. magna	Invert.	21-day reproduction		10	Clubbs and Brooks (2007)
Diphen-hydramine	Antihistamine	L. gibba	Plant	7-day growth		>10,750	Berninger et al. (2011)
		D. magna	Invert.	10-day survival, reproduction	46.1, 3.4	27.8, .08	Berninger et al. (2011)
		D. magna	Invert.	21-day survival		0.12	Meinertz et al. (2010)
		P. promelas	Fish	7-day survival, growth, behavioral feeding rate	836.7, 49.1, 5.6	388.3, 24.5, 2.8	Berninger et al. (2011)
Amlodipine	Antihypertensive	H. vulgaris	Invert.	17-day survival; feeding, bud formation, regeneration	10 Regeneration		Pascoe et al. (2003)
Captopril	Antihypertensive	L. minor	Plant	7-day growth	25,000 (EC_{50})		Cleuvers (2003)
Cyclo-phosphamide	Antineoplastic	D. magna	Invert.	21-day reproduction	100,000	56,000	Grung et al. (2008)
Methotrexate	Antineoplastic	V. fischeri	Bacteria	30-min bioluminescence	1,220,000 (EC_{50})		Henschel et al. (1997)
Tamoxifen	Antineoplastic	B. calyciflorus	Invert.	48-h population growth	250 (EC_{50})		DellaGreca et al. (2007)
		C. dubia	Invert.	7-day population growth	0.81 (EC_{50})		DellaGreca et al. (2007)
Fadrozole	Aromatase inhibitor	P. promelas	Fish	21-day reproduction	2 (LOEC)		Ankley et al. (2002)
		P. promelas	Fish	21-day ovary and testis growth	24.8, 51.7		Panter et al. (2004)
		D. rerio	Fish	38-day VTG, 60-day sex ratio	10 (LOEC)		Andersen et al. (2004)

(continued)

Table 2 (continued)

Compound	Category	Species	Trophic group	Endpoint/Duration	LOEC (μg/L)	NOEC (μg/L)	Reference
Atenolol	Antihypertensive	V. fischeri	Bacteria	30-min bioluminesnece	1,304,000 (EC$_{50}$)		Escher et al. (2006)
		H. vulgaris	Invert.	17-day survival; feeding, bud formation, regeneration	No effect		Pascoe et al. (2003)
		D. magna	Invert.	21-day reproduction 2nd generation	8,900	1,480	Kuster et al. (2010)
		H. azteca	Invert.	14-day growth, reproduction	>8,820	8,820	Kuster et al. (2010)
		P. antipodarum	Invert.	42-day growth, reproduction	>9,450	9,450	Kuster et al. (2010)
		P. promelas	Fish	28-day growth	3,200	1,000	Winter et al. (2008)
		P. promelas	Fish	21-day condition index	3,200	1,000	Winter et al. (2008)
		P. promelas	Fish	21-day reproduction	>10,000	10,000	Winter et al. (2008)
		P. promelas	Fish	4-day, 28-day early life stage	>10,000	10,000, 3,200	Winter et al. (2008)
Metoprolol	Antihypertensive	D. magna	Invert.	9-day growth	12,500	6,150	Dzialowski et al. (2006)
		D. magna	Invert.	9-day reproduction	6,150		Dzialowski et al. (2006)
		D. magna	Invert.	9-day heart rate	3,200		Dzialowski et al. (2006)
		O. mykiss	Fish	21-day pathology	1 (Liver), 20 (Gills)		Triebskorn et al. (2007)
		L. minor	Plant	7-day Growth	>320,000 (EC$_{50}$)		Cleuvers (2003)
Propranolol	Antihypertensive	B. calyciflorus	Invert.	48-h reproduction	180		Ferrari et al. (2003)
		C. dubia	Invert.	7-day reproduction	9		Ferrari et al. (2003)
		C. dubia	Invert.	7-day reproduction	250	125	Huggett et al. (2002)
		D. magna	Invert.	9-day growth	440	220	Dzialowski et al. (2006)
		D. magna	Invert.	9-day reproduction	110	55	Dzialowski et al. (2006)
		D. magna	Invert.	9-day heart rate	55		Dzialowski et al. (2006)
		H. azteca	Invert	7-day reproduction	1		Huggett et al. (2002)

Compound	Classification	Species	Group	Endpoint			Reference
		H. azteca	Invert.	27-day reproduction	100	1	Huggett et al. (2002)
		M. galloprovincialis	Invert.	10-day behavior: feeding rate	147	11	Sole et al. (2010)
		O. latipes	Fish	28-day egg production, deformities	0.5–1 (LOEC), 0.5–1,000		Parrott and Bennie (2009)
		D. magna	Invert.	21-day reproduction	800	400	Stanley et al. (2006)
		P. promelas	Fish	7-day growth	128		Stanley et al. (2006)
		L. minor	Plant	7-day growth	114,000 (EC$_{50}$)		Cleuvers (2003)
R-Propranolol	Antihypertensive	*D. magna*	Invert.	21-day reproduction	800	400	Stanley et al. (2006)
S-Propranolol	Antihypertensive	*D. magna*	Invert.	21-day reproduction	800	400	Stanley et al. (2006)
Etidronic acid	Bone resorption inhibitor	*D. magna*	Invert.	21-day reproduction	>12,000		Gledhill and Feijtel (1996)
Digoxin	Cardiac stimulant	*H. vulgaris*	Invert.	Survival; feeding, bud formation, regeneration	No effect (17-day); 10 Regeneration		Pascoe et al. (2003)
Cotinine (nicotine metabolite)	Cholinergic agonist	*L. gibba*	Plant	7-day growth	>1,000 (EC$_{10}$)		Brain et al. (2004a)
Nicotine	Cholinergic agonist	*D. pulex*	Invert.	21-day reproduction	<7,000		US FDA-CDER (1996)
Iopromide	Diagnostic aid	*V. fischeri*	Bacteria	30-min bioluminescence	$>1.0 \times 10^6$		Vandenbergh et al. (2003)
		P. putida	Bacteria	16-h growth	$>1.0 \times 10^6$		Vandenbergh et al. (2003)
		D. magna	Invert.	21-day reproduction	$>1.0 \times 10^6$		Schweinfurth et al. (1996)
		D. magna	Invert.	22-day reproduction	$>1.0 \times 10^6$		Vandenbergh et al. (2003)
		D. rerio	Fish	28-day hatchability	$>1.0 \times 10^6$		Gyllenhammar et al. (2009)

(continued)

Table 2 (continued)

Compound	Category	Species	Trophic group	Endpoint/Duration	LOEC (µg/L)	NOEC (µg/L)	Reference
Bendroflumethiazide	Diuretic	H. vulgaris	Invert.	17-day survival; feeding, bud formation, regeneration	No effect		Pascoe et al. (2003)
Furosemide	Diuretic	H. vulgaris	Invert.	17-day survival; feeding, bud formation, regeneration	No effect		Pascoe et al. (2003)
17α-Estradiol	Estrogen	O. latipes	Fish	21-day testis–ova Induction	<0.0263	<0.0293	Kang et al. (2002)
	Estrogen	O. latipes	Fish	21-day VTG Induction		0.0293	Kang et al. (2002)
17α-Ethinylestradiol	Estrogen	D. magna	Invert.	Reproduction	105 (EC$_{50}$)	10	Halling-Sorensen et al. (1998)
		C. riparius	Invert.	Mouthpart deformities	0.01		Watts et al. (2003)
		G. pulex	Invert.	Juv. Recruit./sex ratio	0.1		Watts et al. (2002)
		D. magna	Invert.	21-day reproduction	12.5 (EC$_{10}$) 105 (EC$_{50}$)	10	Kopf (1995)
		D. magna	Invert.	21-day reproduction, multigen	0.1 (LOEC)	1,000	Clubbs and Brooks (2007)
		H. azteca	Invert.	35-week multigen	0.1 (LOEC)		Vandenbergh et al. (2003)
		B. tentaculata	Invert.	28-day growth	<0.000125		Belfroid and Leonards (1996)
		L. stagnalis	Invert.	28-day growth	<0.00125		Belfroid and Leonards (1996)

Species	Type	Endpoint	Value	Value	Reference
C. tentans	Invert.	10-day survival, growth	4,100, 1,600 (LC$_{50}$, EC$_{50}$)	6,600, 800 (LC$_{10}$, EC$_{10}$)	Dussault et al. (2008)
H. azteca	Invert.	10-day survival, growth	1100, 160 (LC$_{50}$, EC$_{50}$)	1300, 200 (LC$_{10}$, EC$_{10}$)	Dussault et al. (2008)
M. comuarietis	Invert.	5-month sex developement	<0.001		Schulte-Oehlmann et al. (2004)
N. spinepes	Invert.	24-day reproduction	50		Breitholtz and Bengtsson (2001)
O. latipes	Fish	21-day survival, reproduction, Histo., VTG	0.448, 0.116, 0.063, 0.063		Seki et al. (2002)
P. promelas	Fish	Full life-cycle	0.001		Lange et al. (2001)
P. promelas	Fish	21-day VTG	0.001		Pawlowski et al. (2004)
P. promelas	Fish	21-day testes structure	0.001		Pawlowski et al. (2004)
P. promelas	Fish	21-day liver structure	0.001		Pawlowski et al. (2004)
P. promelas	Fish	21-day fertilization rate	0.01		Pawlowski et al. (2004)
O. mykiss	Fish	Testicular growth	0.02		Jobling et al. (1996)
O. mykiss	Fish	2-month reproduction, sex differentiation, aromatase activity	<0.0001		Purdom et al. (1994)
O. mykiss	Fish	2-month reproduction, sex differentiation, aromatase activity	<0.0003		Sheahan et al. (1994)
D. rerio	Fish	38-day VTG	0.002		Orn et al. (2003)
D. rerio	Fish	7-day reproduction	32		Lister et al. (2009)
R. pipiens	Amph.	Survival, growth, metamorphosis, gonad histology	No effect (Survival, growth); 5 nM inhibited metamorph., altered histology		Hogan et al. (2008)

(continued)

Table 2 (continued)

Compound	Category	Species	Trophic group	Endpoint/Duration	LOEC (µg/L)	NOEC (µg/L)	Reference
17β-Estradiol	Estrogen	N. spinepes	Invert.	24-day reproduction	160 (LOEC)		Breitholtz and Bengtsson (2001)
		O. latipes	Fish	21-day reproduction	0.01 (LOEC)		Hutchinson et al. (2003)
		D. rerio	Fish	78-day survival, behavior, growth, sex develop.	0.0037 nM		Segner et al. (2003a, b)
Diethylstibestrol	Estrogen	D. magna	Invert.	2-day gen growth/ reproduction	62 (LOEC)		Baldwin et al. (1995)
		N. spinepes	Invert.	24-day reproduction	3 (LOEC)		Breitholtz and Bengtsson (2001)
		T. battagliai	Invert.	21-day reproduction	10 (LOEC)		Hutchinson et al. (2003)
		O. latipes	Fish	21-day reproduction	0.01 (LOEC)		Hutchinson et al. (2003)
Drospirenone	Progestogen	P. promelas	Fish	21-day reproduction	6.5		Zeilinger et al. (2009)
Levonorgestrel	Progestogen	P. promelas	Fish	21-day reproduction	0.0008		Zeilinger et al. (2009)
							Pascoe et al. (2003)
							Cleuvers (2003)
							Cleuvers (2003)
Atorvastatin	Lipid lowering	L. gibba	Plant	7-day growth	85 (EC_{10}), 135 (EC_{50})		Brain et al. (2006)
		L. gibba	Plant	7-day growth	300		Brain et al. (2004)
		C. tentans	Invert.	10-day survival, growth	14,300, 10,200 (LC_{50}, EC_{50})	8,200, 5,600 (LC_{10}, EC_{10})	Dussault et al. (2008)
		H. azteca	Invert.	10-day survival, growth	1,500, 2,400 (LC_{50}, EC_{50})	400, 1,400 (LC_{10}, EC_{10})	Dussault et al. (2008)

Bezafibrate	Lipid lowering	H. attenuata	Invert.	96-h morphology, feeding	1,000, 85,900 (EC$_{50}$)	100	Quinn et al. (2008)
		B. calyciflorus	Invert.	48-h population growth	312.5	156	Isidori et al. (2007)
		C. dubia	Invert.	7-day population growth	47	23	Isidori et al. (2007)
Clofibrate	Lipid lowering	D. magna	Invert.	21-day survival; reproduction	8.4 (EC$_{10}$) 106 (EC$_{50}$)	10	Kopf (1995)
		B. calyciflorus	Invert.	48-h reproduction	740	246	Ferrari et al. (2003)
		C. dubia	Invert.	7-day reproduction	2,560	640	Ferrari et al. (2003)
		C. riparius	Invert.	28-day emergence	>8 (mg/kg)		Nentwig et al. (2004)
		L. variegatus	Invert.	28-day reproduction	>8 (mg/kg)		Nentwig et al. (2004)
		D. rerio	Fish	10-day reproduction/ embryo Survival	14,000	70,000	Ferrari et al. (2003)
		P. promelas	Fish	21-day lipid metabolism	108,900		Weston et al. (2009)
		P. promelas	Fish	21-day reproduction	10 (TEC)		Runnells et al. (2007)
		L. minor	Plant	7-day growth	12,500(EC$_{50}$)		Cleuvers (2003)
Clofibric acid	Lipid lowering	V. fischeri	Bacteria	30-min (EC$_{50}$)	100		Henschel et al. (1997)
		V. fischeri	Bacteria	30-min (EC$_{50}$)	91.83		Ferrari et al. (2003)
		T. pyriformis	Invert.	48-h growth (EC$_{50}$)	175,000		Henschel et al. (1997)
		C. dubia	Invert.	7-day reproduction	2,560	640	Ferrari et al. (2003)
		D. rerio	Fish	10-day survival	140,000	70,000	Ferrari et al. (2003)
		O. mykiss	Fish	21-day cytopathology	>100 (Liver and Kidney, 5 (Gill)		Triebskorn et al. (2007)
Fenofibrate	Lipid lowering	L. minor	Plant	7-day growth (EC$_{50}$)	12,500		Cleuvers (2003)
		B. calyciflorus	Invert.	48-h population growth	312.5	156	Isidori et al. (2007)
		C. dubia	Invert.	7-day population growth	39	78	Isidori et al. (2007)

(continued)

Table 2 (continued)

Compound	Category	Species	Trophic group	Endpoint/Duration	LOEC (µg/L)	NOEC (µg/L)	Reference
Gemfibrozil	Lipid lowering	V. fischeri	Bacteria	30-min bioluminescence (EC_{50})	85,740		Isidori et al. (2007)
		V. fischeri	Bacteria	24, 48-h bioluminescence (EC_{50})	64,600, 45,100		Zurita et al. (2007)
		H. attenuata	Invert.	4-day morphology, feeding	1,000	100	Quinn et al. (2008)
		B. calyciflorus	Invert.	48-h population growth	312	156	Isidori et al. (2007)
		C. dubia	Invert.	7-day population growth	156	78	Isidori et al. (2007)
Lovastatin	Lipid lowering	L. gibba	Plant	7-day growth	160 (EC_{50})		Brain et al. (2006)
Caffeine	Stimulant	C. dubia	Invert	7-day growth, mortality	40,000–50,000 (LC_{50})	20,000–50,000 (EC_{25})	97
		P. promelas	Fish	7-day growth, mortality	50,000–60,000 (LC_{50})	30,000–90,000 (EC_{25})	97
		L. gibba	Plant	7-day growth	>1,000 (EC_{10})		Brain et al. (2004b)

Table modified from Crane et al. 2006 and Santos et al. 2010

[a]TEC is toxic effect concentration

In a 21-day full life-cycle reproduction study using *D. longispina* and *D. magna* (Marques et al. 2004), the most toxic compound to invertebrates was the aspirin metabolite, gentisic acid. This compound displayed the lowest value (320 µg/L) for the observed effect concentration (lowest observable effect concentration [LOEC]), in regards to total number of offspring produced. Survival and reproduction have also been monitored for fish (*D. rerio* and *L. machrochirus*); however, fish reproduction and embryo survival in fish after 10-day exposure was not as sensitive as histological endpoints for anti-inflammatory compounds (Ferrari et al. 2003). Digestive tract and gill histology has been examined in *O. mykiss* for the analgesic compound diclofenac. Digestive tract effects are known side effects of NSAIDs in humans (Buttgereit et al. 2001) and other organisms (Dinchuk et al. 1995). Furthermore, digestive tract histology is approximately 100–200× more sensitive as an endpoint than is gill histology over similar time periods (Triebskorn et al. 2004). However, diclofenac is the only analgesic compound for which a digestive tract histology endpoint was measured. Schwaiger et al. (2004) observed alterations in digestive tract histology at concentrations of 5 µg/L (28-day LOEC) in *O. mykiss* for diclofenac, and this has since been observed to be the most sensitive endpoint studied.

Acetaminophen effects on survival, growth, and behavior have also been investigated in two amphibian species, *Rana pipiens* and *Bufo americanus*. *R. pipiens* was not affected after 28-day exposure (Fraker and Smith 2004); however, *B. americanus* experienced significant effects on survival (LOEC of 100 µg/L) (the latter contradicts 28-day exposure for *R. pipiens*) and behavior (LOEC=1 µg/L) after 14-day of exposure (Smith and Burgett 2005).

Ibuprofen effects have been examined in dermatophyte fungi; growth was affected at concentrations >200 mg/L (Sanyal et al. 1993). Chronic toxicity of analgesic compounds to plants has been investigated in duckweed, *L. gibba* and *L. minor*. In general, plants do not appear to be sensitive to analgesic effects. Growth of *L. minor*, in response to 7-day diclofenac and ibuprofen exposures, resulted in EC_{50} values of 7.5 and 22 mg/L, respectively (Cleuvers 2003). In summary, concentrations of analgesics needed to cause toxic effects in wildlife are relatively high. Based on observed environmental concentrations, growth or reproductive effects are not expected to be observed in nature.

3.2 Androgens

Methyltestosterone, which activates androgen response elements, has been observed to induce chronic effects in fish and invertebrates. A 12-week reproduction study with *L. stagnalis* (great pond snail) demonstrated decreased reproduction at 1 µg/L (Czech et al. 2001), and sex development was highly affected at concentrations <0.1 µg/L in a longer study (12 month) performed with *M. cornuarietis* (ramshorn snail) (Schulte-Oehlmann et al. 2004). Twenty-one-day exposure to testosterone has also been observed to reduce the numbers of female *D. magna* produced (Mu and

LeBlanc 2002). Goldfish (*C. carassius*) and killifish (*O. latipes*) appeared to have greater sensitivity than snails when examining reproduction and sex ratio after 80- and 42-day methyltestosterone exposure, respectively (Fujioka 2002; Zerulla et al. 2002). A significantly greater number of males were produced in relation to females at concentrations as low as 0.01 µg/L (Fujioka 2002). Increased production of male fathead minnows (*P. promelas*) was not as sensitive as other endpoints as determined in a shorter phenotypic sex study (13-days during development) (Zerulla et al. 2002). Results of these studies indicate a potential risk in instances of increased environmental concentrations, although limited evidence suggests this is not common.

3.3   Antibiotics

Antibiotics are the most well-studied pharmaceutical compounds, although they also have the greatest variation in mode of action (MOA). As we did with acute effects, antibiotics will be addressed by their general antibiotic class (i.e., fluoroquinolone, β-lactams, macrolides, sulfonamide, tetracycline, etc.), and thus MOA. Macrolide and fluoroquinolone antibiotics are the most studied antibiotic classes. Macrolide antibiotics (ex. erythromycin) inhibit protein synthesis, whereas fluoroquinolone antibiotics (ex. ciprofloxacin) inhibit DNA synthesis and replication through the DNA-gyrase enzyme. The majority of chronic toxicity studies performed with antibiotics have focused on plants and invertebrates; the numerous plant studies likely results from plants having similar target sites as bacteria (Brain et al. 2008a; 2008b). LOEC values for fluoroquinolone antibiotics are highly variable and range between 5 µg/L and 25 mg/L, depending on compound, species, and endpoint investigated (Ferrari et al. 2003, 2004). Sulfonamide antibiotics (ex. sulfadimethoxine) (inhibit folic acid synthesis) are consistently the most toxic to algae, with LOEC values from 5.9 to 1,250 µg/L (Ferrari et al. 2004).

Plant reproduction appears to be a more sensitive endpoint than many other standardized biological ones studied (i.e., growth) to date for antibiotics. LOEC values, based on reproductive effects in *L. minor*, range between 51 and 203 µg/L for all antibiotics except flumequine (fluoroquinolone class antibiotic), which was significantly less toxic (Robinson et al. 2005; Cleuvers 2003). Sulfamethoxazole (sulfonamide antibiotic) is toxic to *L. gibba* at relatively low concentrations. Brain et al. (2008a) provided EC_{10} values of 58.2, 21.2, and 0.655 µg/L for the standardized endpoints of frond number and fresh weight, and the biomarker *para*-aminobenzoic acid, respectively.

Invertebrate reproduction in response to chronic antibiotic exposure has only been investigated in *B. calyciflorus* and *H. vulgaris*. LOEC values are <12.5 and 25 mg/L for fluoroquinolone and sulfonamide antibiotics, respectively (Ferrari et al. 2003). Invertebrates are generally less sensitive to all antibiotic compounds than are plants, which may result from a lack of receptors. No researchers have investigated long-term effects of antibiotics in wastewater for fish or amphibians. Notwithstanding, based on acute and chronic data available for invertebrates, adverse effects on standardized endpoints at environmentally relevant concentrations are not likely.

3.4 Antidepressants

The most prescribed class of antidepressants is the SSRIs. The SSRIs are named for their ability to affect the serotonin reuptake pump on the cell body and dendrites, but not the axon terminals of neurons (Auerbach et al. 1995). Serotonin is important in a number of biologic functions including appetite, immune response, and behavior (Daughton and Ternes 1999; Brooks et al. 2003a), and is also associated with other physiological mechanisms in invertebrates (Santos et al. 2010).

The consequences of SSRI exposure has been studied in invertebrates, fish, amphibians, algae, and plants. Fluoxetine is the most widely studied SSRI to date (Oakes et al. 2010). Although all SSRIs have similar modes of action, particularly in organisms with evolutionary conservation of the SERT (serotonin reuptake transporter) (e.g., fish; Gould et al. 2007), chronic toxicity varies considerably among individual compounds. For aquatic plants, fluoxetine and sertraline did not produce chronic toxicity effects (growth) in duckweed (*L. gibba* and *L. minor*) at the highest concentrations tested thus far (1 mg/L; Brain et al. 2004a). However, among a variety of standardized endpoints examined in green alga, cladoceran, amphipods, insects and fish, algal growth was the most sensitive. This was hypothesized to be due to efflux pump interference (Brooks et al. 2003b). This study highlights the importance of considering unanticipated effects in nontarget organisms (Brooks et al. 2003b), because only algae do not possess the therapeutic receptors targeted by the SSRIs.

The effects of the SSRIs in fish have been investigated in only a handful of studies, and fluoxetine was the tested agent in most of these (Kreke and Dietrich 2008). In most of these studies, the primary focus was on standardized endpoints. Foran et al. (2004) found that a 28-day fluoxetine exposure (100–500 µg/L) caused significantly different estradiol concentrations in fish (*O. latipes*). Development of mosquitofish was also severely inhibited by exposure to 5 µg/L of fluoxetine over their entire life-cycle (Henry and Black 2008). It has been postulated that behavioral endpoints would be highly sensitive to SSRIs due to the effects of antidepressants on the nervous system (Brooks et al. 2003a; Kreke and Dietrich 2008). Studies have demonstrated that feeding behavior of juvenile fathead minnows was more sensitive than growth and mortality endpoints that were employed in standardized 7-day study designs (Stanley et al. 2007; Valenti et al. 2009). Based on these two reports, it appears that the potency of fluoxetine is higher than that of sertraline. Furthermore, lower fluoxetine EC_{10} values were reported for fathead minnow growth and feeding behavior than for similar sertraline thresholds (Stanley et al. 2007; Valenti et al. 2009).

SSRIs have recently been examined for their potential to disrupt the hypothalamus–pituitary–thyroid (HPT) axis in amphibians and fish. Amphibian metamorphosis requires high levels of thyroid hormone (Kollros 1961), and thus, SSRIs could alter metamorphosis in amphibians. Black et al. (2009) and Conners et al. (2009) reported that decreased time was required for *X. laevis* to complete metamorphosis, and decreased body mass at metamorphosis occurred from exposure to sertraline. These results indicate the potential for SSRIs to impact the HPT axis in amphibians, and illustrate the potential for endocrine-related effects in other vertebrates as well. The toxicity of antidepressants is highly variable but is generally less than the actual environmental concentrations. However, SSRIs have been demonstrated to accumulate in laboratory experiments and in field-collected aquatic organisms (Brooks et al. 2005; Chu and Metcalfe 2007; Nakamura et al. 2008; Ramirez et al. 2007, 2009; Kwon et al. 2009; Schultz et al. 2010). Oakes et al. (2010) conducted a risk assessment of fluoxetine and identified a PEC/PNEC >1 (Predicted Environmental Concentration/Predicted No Effect Concentration), indicating a potential risk in surface waters, based on European guidelines (EMEA/CHMP 2006).

Additional considerations should be made for the enantiomer-specific and chirality-specific toxicity of the antidepressants. Stanley et al. (2007) identified that *S*-fluoxetine was up to 9.4-fold more potent to fish than *R*-fluoxetine. This observation mirrored enantiomer-specific therapeutic activity in mammals from the higher potency of the primary metabolite *S*-norfluoxetine. It is also important to note that the SSRIs fluoxetine and sertaline are weak bases that ionize under environmentally realistic conditions; thus, site-specific pH (Kim et al. 2010; Valenti et al. 2011) can strongly influence the aquatic bioavailability and toxicity of the SSRIs (Nakamura et al. 2008; Valenti et al. 2009). Such studies of chirality (Stanley and Brooks 2009) and ionization (Valenti et al. 2011) highlight critical uncertainties presented by the SSRIs in ecological risk assessment.

3.5   Antidiabetics

Metformin is the only antidiabetic drug that has been investigated for chronic toxicity, but was not toxic to the only species tested, *L. minor* (Cleuvers 2003). Metformin likely activates AMP (adenosine monophosphate)-activated protein kinases, thus increasing fatty acid oxidation and glucose uptake (Crane et al. 2006), and thus could cause effects in vertebrate species, although this would not be expected at environmentally realistic concentrations.

3.6   Antiepileptics

Invertebrates (*B. calyciflorus* and *C. dubia*) appear to be most sensitive to carbamazepine exposure, and display decreased reproduction and survival effects at concentrations as low as 25 µg/L (Ferrari et al. 2003). However, of nontraditional endpoints, behavior in *Gammarus pulex* has been the most sensitive, and is more than 60× more sensitive than any other endpoint studied. These results would be expected because carbamazepine blocks voltage-gated Na$^+$ channels in the brain. Therefore, it is likely that behavioral effects will result from carbamezapine exposure. However, few studies have been conducted to examine behavioral effects, and this is a major shortcoming in the carbamezapine database. Emergence of the benthic invertebrate, *Choronomus riparius*, was decreased after a 28-day exposure to 0.625 mg/kg carbamezapine (Nentwing et al. 2004), with other benthic invertebrates (*H. azteca* and *C. tentans*) being relatively insensitive (Dussault et al. 2008). Fish hatching and larval mortality were decreased at concentrations of 25 mg/L in *D. rerio* (Ferrari et al. 2003).

3.7 Antiestrogens

The antiestrogen compound ZM 189,154 competitively inhibits estrogen receptors. Andersen et al. (2004) observed increased amounts of vitellogenin (VTG) in male fish in response to a 38-day exposure of *D. rerio*. VTG is a commonly used biomarker of exposure to endocrine-disrupting compounds for males. VTG is usually only produced in female, oviparous species as a result of high circulating estradiol levels (Larsson et al. 1994). ZM 189,154 exposure also produced undifferentiated gonads in *D. rerio* as a result of a 60-day exposure (Andersen et al. 2004). In a multigenerational study with the selective estrogen receptor modulator faslodex (or, 182,780), Clubbs and Brooks (2007) reported a *D. magna* reproduction no observable effect concentration (NOEC) value of 100 μg/L.

3.8 Antifungals

Azole antifungal drugs, and the aromatase inhibitor fadrozole, are potent, nonspecific inhibitors of cytochrome P450s (CYP) in vertebrates (Ankley et al. 2006). Azoles have been demonstrated to inhibit CYP1A, CYP3A, and CYP19 isoforms in vertebrates, which are vital in xenobiotic and steroid metabolism and the synthesis of estrogen (Hegelund et al. 2004; Hasselberg et al. 2005; Corcoran et al. 2010).

   Fadrozole is the only aromatase inhibitor that has been investigated for potential chronic effects, and has been studied for its reproductive effects. Ankley et al. (2002) and Andersen et al. (2004), respectively, evaluated fadrozole in *P. promelas* and *D. rerio*. Ankley et al. (2007) identified decreased reproduction (LOEC) at a concentration of 2 μg/L in *P. promelas* as a result of 21-day exposure, making it one of the most potent pharmaceuticals studied. Andersen et al. (2004) observed increased VTG and decreased gonad differentiation at slightly higher concentrations in *D. rerio*.

3.9 Antihypertensives

Antihypertensive pharmaceuticals have a wide range of MOAs. They function by inhibiting circulating and tissue angiotensin (captopril; Faxon et al. 1981), binding to benzodiazepine receptors (diazepam; Collinge et al. 1983), or inhibiting calcium ion influx (amlodipine; van Zwieten 1994) to treat cardiac conditions such as angina, heart failure, high blood pressure, and glaucoma (Owen et al. 2007). Very little research has been conducted to examine the chronic effects of antihypertensive medication on aquatic organisms, although studies that have been completed indicate that there is minimal or no toxicity.

Another form of antihypertensive pharmaceuticals is the β-blockers. The β-blockers inhibit β-adrenergic receptors in the heart, either competitively or by antagonizing β1 and β2 adrenergic receptors, and are used to manage cardiac arrhythmias and as a therapy after heart attacks. Fish have similar β-receptors to those that exist in the human heart, liver, and reproductive system, and thus have the potential to be deleteriously affected by exposure (Haider and Baqri 2000; Nickerson et al. 2001). Numerous researchers have investigated the effects of β-blockers on algae, plants, and aquatic invertebrates. However, only a few have examined β-blocker toxicity in fish, although these studies indicate fish were less sensitive than other trophic groups (Parrott and Bennie 2010). In one study, it was found that algal growth was stunted (LOEC) at concentrations as low as 94 μg/L (propanolol) (Ferrari et al. 2004); however, most algal species do not display different growth effects until exposed to β-blockers >1 mg/L (Cleuvers 2003; Ferrari et al. 2004). Propanolol and metoprolol toxicity have been examined in aquatic plants (*L. minor*) and were found to be relatively nontoxic (7-day growth LOEC = 114 and 320 mg/L, respectively) (Cleuvers 2003). For invertebrates, 7-day reproduction in *C. dubia* was highly variable with one study identifying reproductive effects occurring at 9 μg/L (Ferrari et al. 2003), whereas Huggett et al. (2002) did not observe effects until *C. dubia* were exposed to 125 μg/L propanolol. *H. azteca* reproduction was a highly sensitive endpoint for exposure to propanolol (Huggett et al. 2002). Dzialowski et al. (2006) examined the multigenerational responses of *D. magna* to propanolol, reporting that suppression of heart rate and respiration was more sensitive than the standardized endpoint fecundity. Küster et al. (2010) conducted a risk assessment on atenolol and found no unacceptable risk based on European guidelines (EMEA/CHMP [Committee for Medicinal Products for Human Use] 2006).

As noted above for antidepressants, Stanley et al. (2006) compared the toxicity of a racemic mixture of propranolol (*R/S*-propranolol) to each enantiomer individually (*R*-propranolol and *S*-propanolol). Though no differences between enantiomers was observed for *D. magna* reproduction (LOEC = 800 μg/L), the *S*-enantiomer of propanolol, which is more pharmacologically potent in mammals, was more potent to fish. Stanley et al. (2006) observed significantly reduced growth in *P. promelas* as a result of exposure to *S*-propanolol as compared to the *R*-enantiomer (Stanley et al. 2006).

3.10   Bone Resorption Inhibitors

Etidronic acid is a bone resorption inhibitor that functions by reducing osteoclastic activity to prevent bone resorption, and also to prevent bone calcification.

Currently, no effects have been observed on aquatic species of this agent, although *D. magna* has been the only species examined (Gledhill and Feijtel 1992).

3.11 Cholinergic Agonists

The cholinergic agonist nicotine, and its metabolite cotinine, have only been examined in two studies for long-term effects on aquatic organisms. FDA-CDER (1996) examined nicotine effects in *D. pulex* after a 21-day exposure, and identified decreased reproduction at concentrations <7 mg/L. Cotinine growth effects on *L. gibba* were observed at concentrations >1 mg/L (Ankley et al. 2002).

3.12 Diuretics

Two diuretics, bendroflumethiazide and furosemide, have been evaluated for chronic effects in the cnidarian *H. vulgaris*. These two diuretics have different modes of action, with bendroflumethiazide acting by interfering with renal electrolyte reabsorption, and furosemide inhibiting sodium–potassium-chloride transport in the Loop of Henle. After a 17-day exposure, neither compound was observed to have an effect on *H. vulgaris* survival, feeding, bud formation, and regeneration at the highest concentration tested (0.01 mg/L; Pascoe et al. 2003).

3.13 Estrogens and Progestins

Synthetic and endogenous estrogens are hormones that bind to estrogen receptors and cause a wide variety of responses during contraception and development. Estrogens are responsible for numerous physiological functions, such as VTG synthesis and egg yolk protein production in oviparous organisms and gonadal differentiation, development of secondary sex characteristics, and gonadotropin secretion in vertebrates (Larsson et al. 1999).

The clearest evidence of potential adverse effects from chronic exposure to pharmaceuticals is for the synthetic steroid EE2, which elicited effects at low and environmentally relevant concentrations (Lange et al. 2001; Parrott and Blunt 2005). EE2 is the main component used in female birth control, and is the most commonly studied synthetic estrogen. Several authors have investigated chronic toxicity of EE2 to invertebrates. A 35-week multigeneration exposure design was used with *H. azteca* to investigate effects on reproduction (Vandenburgh et al. 2003). Reproduction was significantly reduced as a result of exposure to 0.1 μg/L EE2. Mouthpart deformities (a developmental endpoint) occurred in *C. riparius* after exposure to EE2 at concentrations as low as 0.01 μg/L (Watts et al. 2003); however, other invertebrates (e.g., *H. azteca* and *Chironomus tentans*) have been shown to be relatively insensitive (Dussault et al. 2008). Reproduction and sex ratio have also been examined in invertebrates (*D. magna*, *G. pulex*, *Nitocra spinepes*, *Marisa comuarietis*) exposed to EE2 (Kopf 1995; Halling-Sorensen et al. 1998; Watts et al. 2002). For example, in a multigenerational *D. magna* study with EE2, daphnia fecundity and the biochemical biomarkers associated with endocrine function and population growth rate were not affected at levels up to 1 mg/L. As suggested by Hutchinson (2002), and illustrated by Clubbs and Brooks (2007), vertebrate estrogen agonists and antagonists do not appear to elicit chronic toxicity in aquatic invertebrates at environmental concentrations.

The effects of chronic EE2 exposure have been studied in three different fish species. *P. promelas* was highly sensitive to long-term exposure with a LOEC value of 0.001 µg/L for reproduction (Lange et al. 2001). Other fish species (*O. latipes* and *O. mykiss*) demonstrate similar sensitivities (Jobling et al. 1996; Hutchinson et al. 2003). Additionally, chronic EE2 exposure in amphibians has been investigated for effects on survival, growth, metamorphosis, and gonadal histology. Hogan et al. (2008) found no effect of 5 nM EE2 exposure on *R. pipiens* (for Gosner stage 27–42) survival, weight, or snout–vent length. These researchers did, however, find that EE2-exposed tadpoles took significantly longer to metamorphose, and significantly more of the exposed tadpoles were intersex than were controls, indicating potential endocrine-related effects.

Other estrogens, such as 17β-estradiol and DES have also been investigated for chronic toxicity (e.g., Baldwin et al. 1995; Breitholtz and Bengtsson 2001; Hutchinson et al. 2003). For example, 17β-estradiol caused decreased reproduction in *O. latipes* after a 21-day exposure to 0.01 µg/L (Hutchinson et al. 2003), and also decreased reproduction in the copepod *N. spinepes* at concentrations of 160 µg/L (Breitholtz and Bengtsson 2001). Furthermore, reproductive effects have been identified in numerous studies on invertebrates and fish from DES exposure. In all cases, DES exposure at concentrations of 10 µg/L and below caused significant reductions in reproduction in *D. magna*, *N. spinepas*, *Tisbe battagliai*, and *O. latipes* (Baldwin et al. 1995; Breitholtz and Bengtsson 2001; Hutchinson et al. 2003).

In addition to synthetic estrogens, gestagens are also used for oral contraception and hormone replacement therapy. Levonorgestrel and drospirenone exposure inhibited reproduction and altered gonad histology of *P. promelas* at concentrations of 0.8 ng/L levonorgestrol and 6.5 µg/L drospirenone after a 21-day exposure (Zeilinger et al. 2009). In addition to being highly potent, Fick et al. (2010) found that caged trout from several Swedish rivers that received municipal effluent discharges accumulated levonorgestrol to plasma levels exceeding human therapeutic thresholds (e.g., human maximum plasma concentration [C_{max}]). When coupled with the Zeilinger et al. (2009) results, it appears that progestins may present a significant risk to aquatic life, particularly in effluent-dominated and dependent streams (Brooks et al. 2009a, Daughton and Brooks 2011).

Similar to the androgen methyltestosterone, estrogens are capable of producing effects at low concentrations that are environmentally relevant and possible. This potential has been recognized by numerous researchers. Kidd et al. (2007) conducted a multiyear, whole lake study in which they examined the impacts of synthetic estrogens on a lake system. This was the first known study in which pharmaceutical effects were evaluated under a field scenario. By the second year of the study, these authors found a complete crash of fathead minnow populations caused by a lack of reproduction from feminization of male fish, increased VTG production in male and female fish, altered oogenesis in females, and intersex males (Kidd et al. 2007). Additionally, estrogens have been observed to significantly reduce agonistic behavior in male fish, thus leading to decreased reproductive fitness and reduced reproductive success (Martinovic et al. 2007). Decreased aggressiveness in male fish toward other males, decreased interest in female fish, and decreased agonistic behaviors has been found in other studies as well (Bell 2001; Majewski et al. 2002). These alterations have been observed to be sufficient to cause decreased reproductive fitness and could cause population effects (Martinovic et al. 2007).

3.14 Lipid-Lowering Drugs

Two different types of blood lipid-lowering drugs are currently prescribed, statins and fibrates. Fibrates function by activating lipoprotein lipase and are more commonly studied, and more commonly observed in the environment than are the statins (Staels et al. 1998). Clofibrate, or clofibric acid, causes induction of lipoprotein breakdown, fatty acid reuptake, and reduction of triglyceride production. Chronic toxicity of clofibrate and clofibrinic acid has been investigated in a number of different trophic groups, including benthic and pelagic invertebrates, fish, algae, and plants. Invertebrates appear to be the most sensitive trophic group, with $D.$ $magna$ being the most sensitive invertebrate. $D.$ $magna$ reproduction during a 21-day life-cycle assessment was significantly decreased (EC_{10}) at concentrations of 8.4 μg/L (Kopf 1995). Reproductive endpoints have also been examined for other invertebrates, although they were not as sensitive as $D.$ $magna$ (Ferrari et al. 2003). Fish ($D.$ $rerio$) reproduction and embryo-survival studies indicated clofibrate and clofibrinic acid cause reproductive effects, but only at concentrations 1,000× higher (less toxic) than invertebrates (Ferrari et al. 2003). Additionally, effects of clofibric acid and bezafibrate were investigated for effects on lipid metabolism in $P.$ $promelas$; however, effects were not observed at concentrations lower than 108.9 mg/L (Weston et al. 2009). Runnalls et al. (2007) observed clofibric acid inhibits spermatogenesis in fathead minnows at concentrations much lower (1 mg/L) than any other endpoint studied to date for fish, but still 100× greater than concentrations needed to cause effects in $D.$ $magna$. These data have been supported by Mimeault et al. (2005), who observed reduced plasma testosterone concentrations in goldfish, possibly indicating fibrates may affect androgen synthesis.

For aquatic plants, duckweed growth (7-days) was inhibited (EC_{50}) by clofibrate concentrations of 12.5 mg/L and above (Cleuvers 2003). In addition to fibrates, the statin form of lipid regulators such as atorvastatin has also been examined for chronic growth effects. The statins inhibit cholesterol synthesis by inhibiting the 3-hydroxymethylglutaril coenzyme A reductase enzyme (Laufs and Liao 1998).

Plants are also susceptible to the statins as they share a common enzyme that regulates the mevalonic acid pathway (Disch et al. 1998). Decreased growth in *L. gibba* was observed at lower concentrations than those for clofibrate (Brain et al. 2004a); however, *H. azteca* and *C. tentans* appeared less sensitive to avorstatin than clofibrate (Dussault et al. 2008). Another statin, simvastatin, has been investigated in three different studies (Dahl et al. 2006; DeLorenzo and Fleming 2008; Key et al. 2008); however, in all of these studies, saltwater species were used, which falls outside the scope of this review.

3.15 Other Compounds

Additional compounds that have been investigated for chronic toxicity include the stimulant caffeine, the X-ray medium iopromide and the antihistamine diphenyhydramine. Chronic exposure to caffeine in *C. dubia*, *P. promelas*, and *L. gibba* has not resulted in any observed effects at concentrations below 1 mg/L (Brain et al. 2004a; Moore et al. 2008). The X-ray medium iopromide produced no survival, growth, or reproductive effects in *D. magna* at the highest concentration studied (1,000 mg/L; Schweinfurth et al. 1996).

As introduced above, Berninger et al. (2011) examined the sublethal responses of several model aquatic organisms to the antihistamine diphenhydramine. Though diphenhydramine treatment of over 10 mg/L had no effect on *L. gibba* frond number, wet weight, or growth rate, *P. promelas* and *D. magna* were much more sensitive. The *P. promelas* NOECs (7-days) for survival, growth and feeding behavior were 388.3, 24.5, and 2.8 µg/L, respectively (Berninger et al. 2011). Feeding behavior responses were also more sensitive than the standardized growth endpoint for the SSRIs fluoxetine (Stanley et al. 2007) and sertraline (Valenti et al. 2009), which share a common MOA with diphenhydramine. Interestingly, for this compound, *D. magna* were actually more sensitive than *P. promelas* with survival and reproduction NOECs reported at 27.8 and 0.8 µg/L, respectively (Berninger et al. 2011).

4   Acute to Chronic Ratios and Modes of Action of Human-Use Pharmaceuticals

Ankley et al. (2007) suggested identifying the biological effects of long-term exposures to aquatic organisms based on their pharmaceutical MOA. Therefore, a priori knowledge of MOA and/or potential side effects can assist in identifying: (1) compounds that are most likely to cause adverse effects in aquatic organisms, based on having a similar receptor site to what exists in humans; (2) trophic groups that may be the most sensitive; and (3) the biological (measurement) endpoints that require monitoring during testing. This approach will then lead to a "pharmaceutical-specific" testing program and, therefore, streamline toxicity testing procedures.

Berninger and Brooks (2010) highlighted the importance of MOA by examining the potency of mammalian pharmaceuticals using probabilistic hazard assessment approaches. Assessing the ecological risk of pharmaceuticals through extensive toxicity testing for each drug is an impractical use of resources, time, and animals. Instead, probabilistic models can be employed to determine the probability of finding compounds or endpoint values below specific thresholds. One such probabilistic modeling approach is to use chemical toxicity distributions. These have been utilized to predict antibiotic concentrations of concern in model aquatic plant species (Brain et al. 2006a, b), to compare the sensitivity of various in vitro and in vivo models of estrogenic activity (Dobbins et al. 2008), and to predict the acute and chronic toxicity of parabens to fish and cladocerans (Dobbins et al. 2009). Probabilistic Pharmaceutical Distributions (PPDs), a term coined for chemical toxicity distributions performed on pharmaceuticals that comprise multiple chemical classes and modes of action, has recently been proposed as a screening approach for identifying pharmaceuticals that have potential ecological risk.

Another approach would be to use acute to chronic ratios (ACR). As noted by Rand (1995), the relationship between acute and chronic toxicity, often expressed by the ACR, can provide a useful diagnostic tool to differentiate among modes of action. In the case of pharmaceuticals (e.g., hormones), ACR values may be orders of magnitude higher than those typically observed for historical contaminants (Ankley et al. 2006). This is because pharmaceuticals are designed to be highly selective and to affect specific pathways at low concentrations, which then can produce higher than typical ACR values. Specifically, Berninger and Brooks (2010) identified a statistically significant relationship between fish ACR values and mammalian acute toxicity to therapeutic ratio (similar to margin of safety), but only when a chronic response in fish was potentially linked to the therapeutic MOA. A shortcoming of the Berninger and Brooks (2010) approach was that very few chronic studies in fish have employed measures of effect that relate to drug MOA; thus, multiple endpoints (e.g., fish reproduction, fish behavior) and model organisms were explored in their initial study. The foregoing shows the importance of considering MOA and carefully selecting chronic endpoints while performing environmental assessments of pharmaceuticals (Ankley et al. 2006, 2007).

Potency is another key aspect to examine when determining potential effects in aquatic systems. To illustrate the importance of relative potency of pharmaceuticals (PPDs), PPDs were created to compare human maximum plasma concentration (C_{max}) and rat oral acute toxicity (LD_{50}) for four pharmaceutical classes: antidepressants, antihistamines, reproductive hormones, and NSAIDs (Fig. 1). Details of the experimental mammalian drug database and specific method of PPD development are described in more detail elsewhere (Berninger and Brooks 2010). Briefly, data were numerically ranked and converted into a probability percentage using the Weibull formula. C_{max} and LD_{50} data were then plotted against their respective probability ranks on a log-probability scale. Centiles were calculated using regression analysis. For example, the 5th centile would indicate that 95% of chemicals within that pharmaceutical class would be expected to have a LD_{50} value greater than the 5th centile.

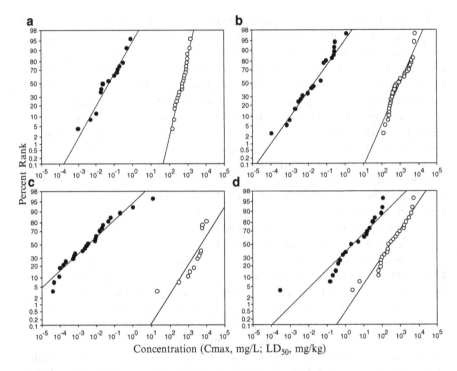

Fig. 1 Probabilistic Pharmaceutical Distribution (PPD) of maximum plasma concentration (C_{max} (·)) and rat acute toxicity (LD$_{50}$, oral (○)) for four classes of pharmaceuticals: (**a**) antidepressants, (**b**) antihistamines, (**c**) reproductive hormones, and (**d**) nonsteroidal antiinflamatory drugs

Pharmaceuticals exhibiting the lowest C_{max} concentrations (and therefore having the greatest potencies) are predicted to require lower environmental exposure concentrations to elicit a pharmacological response. Differences in potencies across pharmaceutical classes can be observed by comparing 5th centiles of the PPD regression line (Table 3). The relative difference between C_{max} and LD$_{50}$ values can also be examined as a screening approach to potentially identify classes of therapeutics that present a greater relative hazard to aquatic vertebrates. A narrow difference between the C_{max} and LD$_{50}$ distributions can indicate chemicals that have the smallest margins of safety, as is observed with NSAIDs. The C_{max} and LD$_{50}$ for NSAIDs are separated by only two orders of magnitude (Fig. 1). Potencies ranked as follows: reproductive   hormones > antihistamines > antidepressants > NSAIDs (Fig. 1; Table 3). This analysis predicts that reproductive hormones should elicit MOA-related environmental effects at lower concentrations than other pharmaceutical classes.

As noted above, efforts have begun in a few studies to examine the aquatic toxicity of different pharmaceutical compounds by using relevant endpoints that are based on pharmaceutical MOA (Table 4). MOA studies have focused on vertebrates (i.e., fish and amphibians), presumably because they have similar targets as do

Table 3 Equation for regression lines and values corresponding to centile values of probabilistic pharmaceutical distributions of maximum plasma concentration (C_{max}; mg/L) and rat acute toxicity (LD_{50}, oral; mg/kg) for four classes of pharmaceuticals: antidepressants (AD), antihistamines (AH), reproductive hormones (RH), NSAIDs (NS)

Compound	Distribution	n	r^2	a	b	Centile value					
						1%	5%	10%	50%	95%	99%
AD	C_{max}	23	0.94	1.249	1.602	7.14×10^{-4}	2.51×10^{-3}	4.90×10^{3}	5.21×10^{-2}	0.554	3.80
	LD_{50}	23	0.97	2.990	-8.021	80.1	135.4	179.0	480.1	1703	2877
AH	C_{max}	36	0.95	0.990	1.640	9.90×10^{-5}	4.83×10^{-4}	1.12×10^{-3}	2.21×10^{-2}	1.01	4.94
	LD_{50}	36	0.93	1.595	-4.763	33.7	90.1	152.3	968	10402	27820
RH	C_{max}	26	0.95	0.652	1.594	9.78×10^{-7}	1.08×10^{-5}	3.91×10^{-5}	3.60×10^{-3}	1.20	13.3
	LD_{50}	26	0.68	1.117	-4.145	42.5	173.1	366.0	5139	152622	621993
NS	C_{max}	25	0.88	0.701	-0.295	1.26×10^{-3}	1.18×10^{-2}	3.91×10^{-2}	2.63	586	5500
	LD_{50}	25	0.95	1.052	-2.586	1.77	7.85	17.4	287.4	10526	46793

n = number of compounds in distribution, a = slope of regression line, b = y-intercept of regression line

Table 4 Examples of studies incorporating mode of action or known side-effects in chronic toxicity studies for human pharmaceuticals

Chemical	Class	Mode of action	Species	Endpoint	References
Diclofenac	Analgesic	Inhibition of cyclooxygenase (COX) by inhibiting prostaglandin synthesis	*Oncorhynchus mykiss*	Liver/gill histology	Triebskorn et al. (2004)
			O. mykiss	Digestive tract histopathology	Schweiger et al. (2004), Mehinto et al. (2010)
			O. mykiss	Kidney necrosis	Mehinto et al. (2010)
Methyl testosterone	Androgen	Activation of androgen response elements	*Carassius carassius*	Phenotypic sex ratio	Fujioka (2002)
			Pimphales promelas	Phenotypic sex ratio	Zerulla et al. (2002)
			Oryzias latipes	Sex reversal	Zerulla et al. (2002)
Fluoxetine	Antidepressant	Inhibition of serotonin reuptake transporter (SERT) on cell body and dendrites	*O. latipes*	Estradiol conc.	Huggett et al. (2003), Foran et al. (2004)
			Gambusia affinis	Sex ratio	Henry and Black (2008)
			P. promelas	Swimming velocity, Predator avoidance	Painter et al. (2009)
			P. promelas	Feeding behavior	Stanley et al. (2007)
			Xenopus laevis	Metamorphosis	Conners et al. (2009)
Sertraline	Antidepressant	Inhibition of serotonin reuptake transporter (SERT) on cell body and dendrites	*P. promelas*	Swimming velocity, Predator avoidance	Painter et al. (2009)
			P. promelas	Feeding behavior	Valenti et al. (2009)
			X. laevis	Metamorphosis	Conners et al. (2009)
ZM 189, 154	Antiestrogen	Competitive inhibition of estrogen receptor	*Danio rerio*	Sex ratio, VTG[a]	Andersen et al. (2004)
Diphenhydramine	Antihistamine, antidepressant anticholinergic	Inhibition of H1 histamine receptor, SERT, aceytlcholine receptor	*P. promelas*	Feeding behavior	Berninger et al. (2011)

Compound	Class	Mode of action	Species	Endpoint	Reference
Fadrozole	Aromatase inhibitor	Inhibition of aromatase enzyme	*D. rerio*	Sex ratio, VTG	Andersen et al. (2004)
Diazepam	Benzodiazapine	GABAergic agonist	*D. rerio*	Swimming behavior, circadian rhythm, gene expression	Zeilinger et al. (2009)
Bezafibrate	Lipid lowering	Induction of lipoprotein breakdown, fatty acid reuptake and reduction of triglyceride production	*P. promelas*	Lipid metabolism	Weston et al. (2009)
Clofibrate	Lipid lowering		*P. promelas*	Lipid metabolism	Weston et al. (2009)
17α-ethinylestradiol	Estrogen	Activation of estrogen receptors	*O. latipes*	Gonad histology, VTG	Seki et al. (2002)
			O. mykiss	Testicular growth	Jobling et al. (1996)
			O. mykiss	Sexual differentiation, Aromatase activity	Purdom et al. (1994)
			R. pipiens	Metamorphosis, gonad histology	Hogan et al. (2008)
17β-Estradiol	Estrogen	Inhibition of ovulation	*D. rerio*	Sexual development	Segner et al. (2003)
Diethylstilbestrol	Estrogen	Inhibition of ovulation	*O. latipes*	Reproduction	Hutchinson et al. (2003)
Drospirenone	Progestogen	Inhibition of ovulation	*P. promelas*	Reproduction	Zeilinger et al. (2009)
Levonorgestrel	Progestogen	Inhibition of ovulation	*P. promelas*	Reproduction	Zeilinger et al. (2009)

[a]Vitellogenin concentration

humans; however, several MOA studies have been performed in other organisms as well. As mentioned, Brain et al. (2008a) proposed the development of a MOA-related biomarker of effect when *L. gibba* were exposure to the sulfonamide antibiotic sulfamethoxazole. In fact, *p*-aminobenzoic acid concentrations, which serve as a precursor for folate synthesis in plants, was identified as a biomarker of effect that was 20× and 40× more sensitive than the standardized endpoints fresh weight and frond number, respectively (Brain et al. 2008b). However, in general, for acute responses, pharmaceuticals typically cause nonspecific narcosis in nonvertebrate species (Cleuvers 2003;Brooks et al. 2008).

The majority of MOA studies have focused on hormones, hormone mimics, or hormone antagonists. In these studies, physiological, histological, and molecular endpoints have been measured. To date, VTG synthesis in sexually mature male fish, following exposure to estrogen agonists, appears to be the most sensitive endpoint; however, VTG synthesis in adult males has not been conclusively linked to individual or population-level effects. This lack of observed effect thus decreases its usefulness as a measure of effect (Sumpter and Jobling 1995) in ecological risk assessment (Ankley et al. 2007) and its use from a regulatory perspective, although it has been suggested in draft regulatory guidelines (OECD 1999; Huet 2000). However, in the case of VTG inhibitors, one interesting study by Miller et al. (2007) demonstrated a relationship between VTG levels in adult female fathead minnows and population growth dynamics. This suggests that it has utility as a biomarker of effect for antiestrogens, although more research is needed to fully assess VTG's usefulness in assessing risk. Endpoints that have been investigated that are indicative of adverse effects on individuals or populations have included sexual differentiation, and histological effects, such as gonadal structure. Other sensitive endpoints in fish have been the production of sperm through spermatogenesis and circulating testosterone levels, in response to estrogenic compound exposure. Again, identifying specific responses as indicators of exposure or measures of effect is critical for integrating such information in risk assessment (Hutchinson et al. 2006). As noted above, Ankley et al. (2007) specifically provided a framework for determining whether sublethal endpoints and biomarkers are indicators of exposure or effect.

Besides hormones and hormonal antagonists, lipid regulating drugs are the only compounds that have been investigated for toxicity that is based on MOA. In a single study, the activation of FAO (fatty acid β-oxidase) was investigated as an indicator of lipid metabolism in *P. promelas* after exposure to bezafibrate and clofibric acid (Weston et al. 2009). FAO is important in β-oxidation of fatty acids in peroxisomes, and functions to convert low density lipoproteins to higher density lipoproteins for fibrate-type lipid regulators (Staels et al. 1998). In contrast to studies with hormones, MOA studies that addressed FAO activation were not as sensitive as other endpoints that were investigated for lipid regulators. Studies in which endpoints related to potential side-effects of lipid regulating drugs were evaluated, i.e., inhibition of androgen synthesis, were more sensitive than MOA endpoints.

The analgesic diclofenac has been investigated in two studies in which the liver, gill, and digestive tract histopathology were examined in rainbow trout; each were found to be highly sensitive endpoints (Schwaiger et al. 2004; Triebskorn et al. 2004). Diclofenac, and other NSAIDs, are known to cause alterations in renal physiology and function in mammals (Revai and Harmos 1999; Manocha and Venkataraman 2000; Hoeger et al. 2005; Schmitt et al. 2010). Perhaps the most well-known and highly publicized effect of diclofenac was on Oriental white-backed vultures on the Indian subcontinent. Oaks et al. (2004) observed that vultures exposed to diclofenac from consuming treated livestock produced renal failure and visceral gout leading to death. Although kidney and digestive tract histology does not reflect the therapeutic MOA of the NSAIDs, this example demonstrates how knowledge of MOA, and potential side-effects, can help identify sensitive endpoints and possible effects in aquatic and nonaquatic species. Mehinto et al. (2010) recently demonstrated the importance of NSAID exposure at environmentally relevant levels to fish. Specifically, kidney necrosis was observed to occur in rainbow trout following a 21-day exposure to 1 µg/L of the NSAID diclofenac (Mehinto et al. 2010). Additionally, it is well known that prostaglandins are needed for bird eggshells, and diclofenac has been observed to cause egg shell thinning (Lundholm 1997), which can result in decreased hatchability. Therefore, based on similarities between human and fish cyclooxygensase a potential sensitive endpoint could be egg production and hatchability in fish exposed to NSAIDs.

Knowledge of MOA and the potential for the interaction among compounds is also important for the antidepressants fluoxetine and sertraline (Kreke and Dietrich 2008). Both agents are suspected to interfere with the HPT axis, either by impeding release of thyrotropin-releasing hormone or by increasing amounts of triiodthyronine in extrathyroidal tissue (Kirkegaard and Faber 1998; Eravci et al. 2000). Amphibian metamorphosis is thyroid-hormone dependent (Kollros 1961), making frogs an ideal species to study the potential endocrine effects of fluoxetine and sertraline. Conners et al. (2009) investigated the metamorphosis of *X. laevis*, and identified effects resulting from exposure to fluoxetine and sertaline. These authors observed that antidepressants caused tadpoles to metamorphose faster than nonexposed tadpoles; however, the tadpoles metamorphosed at a smaller size, which put them at a competitive disadvantage over larger metamorphs (Denver 1997). Results of this study do not support the hypothesis that fluoxetine and sertraline can interfere with the HPT in amphibians. The results do indicate that food intake was dramatically reduced, likely from increased levels of serotonin, which has been demonstrated to decrease food intake and foraging in many species (Meguid et al. 2000; Crespi and Denver 2004). As reported by several researchers, the SSRIs are also recognized to be endocrine disruptors and modulators in fish (Brooks et al. 2003a; Foran et al. 2004; Lister et al. 2009; Mennigen et al. 2010a, b; Oakes et al. 2010; Mennigen et al. 2011).

In addition to sexual-related endpoints (i.e., VTG, gonad histology, sperm production), competitive behavior should be examined, because it is highly sensitive and directly relates to reproductive fitness (Martinovic et al. 2007). Flouxetine has been observed to cause stimulation of spawning and reproduction in invertebrates, probably as a result of increased synaptic serotonin levels, which can stimulate ecdysteroids, ecdysone, and juvenile hormone (LeBlanc et al. 1999; Nation 2002). Flouxetine also causes the release of gonadotropin in fish which is responsible for sex steroid synthesis and controls oogenesis (Arcand-Hoy and Benson 2001). An additional microcosm study was performed, in which the SSRI effects on zooplankton communities was investigated (Laird et al. 2007). After 35-day of exposure, only copepod abundance and copepod species richness was affected by high concentrations of SSRIs (Laird et al. 2007). Stimulatory effects were observed for all zooplankton studied (Rotifers, Clodocerans, Copepods) after short-term exposures (2–8-days), although the stimulatory effects disappeared thereafter (Laird et al. 2007). SSRIs have been observed to alter swimming, social, and feeding behavior in fish. Decreased swimming in Chinook salmon (*Oncorhynchus tshawtscha*) and mosquitofish (*G. affinis*) were observed in response to fluoxetine exposure (Clements and Schreck 2007; Henry and Black 2008). Decreased aggression in male fish probably from increased serotonin levels has also been observed for *Thalassoma bifasciatum* (Bluehead wrasse) as a result of 14-day exposure to fluoxetine (Perreault et al. 2003; Semsar et al. 2004; Clotfelter et al. 2007).

Decreased feeding activity has also been observed for hybrid striped bass and *P. promelas* as a result of fluoxetine exposure (Stanley et al. 2007; Gaworecki and Klaine 2008). As noted above, sertraline has been shown to decrease feeding activity in *P. promelas* (Valenti et al. 2009); both sertraline and fluoxetine elicited ecologically important behavioral perturbations in juvenile fathead minnows at levels lower than thresholds for standardized endpoints (e.g., mortality, growth). Painter et al. (2009) reported that predator avoidance behavior in juvenile fathead minnows were reduced by environmentally relevant, albeit at relatively high levels, mixtures of SSRIs and the selective norepinephrine reuptake inhibitors venlafaxine and buproprion. Unfortunately, exposure methods among these studies varied, and, therefore future assessments of behavioral responses should be examined to compare sensitivities and potential utility as regulatory toxicity tests for SSRIs and other therapeutics. For example, Nassef et al. (2010) observed increased time-to-feed in fish after a 72-h exposure to diclofenac and carbamezapine, and also observed that carbamezapine decreased swim speed. The concentrations required to cause alterations in behavior for diclofenac and carbamezapine were greater than what was needed to cause histopathological effects, but behavior was much more sensitive than typical endpoints such as growth and development.

A review of MOA-based toxicity testing reveals a lack of studies incorporating these methods and endpoints. Pharmaceutical compounds such as antiepileptics could alter behavior from action in the brain, and antidiabetic pharmaceuticals could alter renal physiology and function; however, to date, these potential endpoints have been examined in only a few studies. MOA studies are extremely important in identifying the potential effects of pharmaceuticals, because these effects are likely the most sensitive endpoint, as observed for hormones. MOA studies can be used to help identify sensitive endpoints that will help define more clearly what ecological effects exist, and what the potential risks of pharmaceuticals that appear in wastewater effluent and surface water are.

5 In Vitro Assays and Computational Toxicology

In vitro assays and computational pharmacology and toxicology approaches have the potential to add valuable insights into the potential effects and mechanisms of toxicity caused by pharmaceutical compounds. In addition, these technologies may be useful for prioritizing therapeutics for future studies (Ankley et al. 2007). For example, Escher et al. (2005) provided perhaps the first refereed study that employed a battery of in vitro assays for a targeted group of pharmaceuticals. These authors highlighted the utility of employing such approaches to identify compounds that acting through specific modes of action.

Fish cells have been used as a model system for the toxicological testing of many xenobiotic compounds (Powers 1989). In vitro cytotoxic assays using fish cell lines have been used to identify the potential acute and chronic effects of numerous xenobiotics (e.g., Babich and Borenfreund 1987; Saito et al. 1991; Bruschweiler et al. 1995). Caminada et al. (2006) provided an inclusive study in which the pharmaceutical effects of 34 different drugs were evaluated on fish cell lines PLHC-1 (hepatocellular carcinoma cell line) and RTG-2 (rainbow trout fibroblast-like cells). Study results correlated well with the toxicity of the pharmaceuticals for *Daphnia* spp. Henschel et al. (1997) and Laville et al. (2004) found similar results to those of Caminada et al. (2006) for fish cell lines PRTH (rainbow trout hepatocytes), PLHC-1, and BF-2 (bluegill fry) cells, in response to pharmaceutical exposure. Schnell et al. (2009) examined the toxicity of 11 pharmaceuticals in the fish cell line RTL-W1 (rainbow trout liver), and observed that the antidepressants were the most toxic, and the anti-inflammatory agents were the least toxic. Based on these studies, it appears that fish-cell line in vitro assays are very good at predicting the toxicity of a wide range of pharmaceutical compounds in *D. magna*; however, in vitro assay results do not necessarily reflect pharmaceutical toxicity in other species. At a minimum, screening or short-term in vitro studies can produce a preliminary assessment of the potential toxic effects of pharmaceuticals in aquatic organisms. Additionally, the EPA ToxCast program, including Tox21 efforts, has been used to perform a variety of in vitro assays and to develop computational approaches to identify toxicity pathways and support prioritization schemes for thousands of industrial chemicals (http://www.epa.gov/ncct/toxcast/).

A number of other in vitro assays, such as microarrays, comet arrays, and micronucleus assays, are currently used for identifying cellular response and genotoxic effects from xenobiotic exposure. Microarrays were first described in 1995 (Schena et al. 1995) as a means to simultaneously analyze the expression of a large number of genes, in response to toxicant exposure. DNA microarrays are glass slides or membranes that have short DNA sequences (oligonucleotides) fixed to the slide by a covalent bond. These DNA oligonucleotide sequences are characteristic of a specific gene and are used to detect changes in DNA expression levels. mRNA can then be added to the microarray and allowed to hybridize to the slide, providing a gene profile for up- and down-regulated genes, in response to xenobiotic exposure. This profile can therefore indicate toxicant-induced gene expression and potential toxicant MOA (Neuman and Galvez 2002). cDNA microarrays have recently been used to identify the effects of neuro-active pharmaceuticals on the brain activity of zebrafish (van der Ven et al. 2005). These authors identified the brain-specific cDNA microarrays that were sensitive to chlorpromazine (a dopamine antagonist), and demonstrated their potential for use in MOA-based toxicity testing. Larkin et al. (2003) also used cDNA microarrays to investigate the effects of synthetic estrogens on *C. variegatus*. These cDNA microarrays detected dose-dependent alterations in gene expression as a result of exposure to 17β-estradiol, EE2, and DES. These studies indicate the potential that gene expression profiles have to help identify the pharmaceutical effects that would otherwise be difficult to assess by using traditional endpoints (Ankley et al. 2007; Brooks et al. 2008).

Comet assays are another common in vitro technique that can be used to detect DNA damage resulting from xenobiotic exposure, and they can be used for identifying the genotoxicity of xenobiotics (Al-Sabti and Metcalfe 1995; Singh et al. 1998). Comet assays, or single cell gel electrophoresis, use cells that are lysed, stained, suspended in agarose gel, and are then subjected to electrophoresis; the broken fragments move farther during electrophoresis than do the larger segments (Singh et al. 1998). To date, comet assays have not been incorporated into pharmaceutical testing; however, they have been used to identify strand breaks in invertebrates, fish, and amphibians, from exposure to a diverse group of xenobiotics; results show that comet assays have the potential to detect aquatic effects from exposure to pharmaceuticals (Ralph et al. 1996; Mitchelmore et al. 1997; Mitchelmore and Chipman 1998a, b). Micronucleus assays have also been observed to be sensitive in identifying chromosomal damage and genotoxic effects as a result of xenobiotic exposure (Heddle et al. 1991; Masuda et al. 2004). Although they are similar to comet assays, micronucleus assays have not been used to identify genotoxic effects of pharmaceuticals in WWTP effluents, but have been used to identify effects to a broad range of chemicals (Al-Sabti and Metcalfe 1995; Grisolia 2002). Results of these studies indicate that in vitro techniques are valuable for identifying effects (i.e., chromosomal aberration) that are not typically observed by using traditional chronic toxicity testing designs and endpoints. Additionally, in vitro assays and genomic data have practical uses in a tiered testing framework in which genomic approaches can provide large amounts of data in a short-time period (Ankley et al. 2006). Ankley and coauthors (2006) illustrate how use of genomic approaches can support a typical tiered testing framework, which should be useful for evaluating the environmental effects of pharmaceuticals (Ankley et al. 2007; Brooks et al. 2008, 2009b).

Another valuable tool for identifying the potential effects of pharmaceutical exposure in freshwater ecosystems is quantitative structure activity relationships (QSARs). QSARs have been used historically for many industrial chemicals (Escher and Hermens 2002), including several recent studies with pharmaceuticals (Sanderson et al. 2003, 2004; Sanderson and Thomsen 2007, 2009). Future research is needed to integrate next generation computational pharmacology and toxicology tools (e.g., molecular docking) into studies performed to determine the environmental effects of pharmaceuticals (Brooks et al. 2009b).

6 Considerations for Toxicological Testing

When considering how to test human pharmaceuticals that appear in wastewaters and surface waters for toxicity, three points should be carefully considered: (1) what pharmaceutical compound(s) are present, (2) what species are potentially exposed and are sensitive native species among them, and (3) what are the most sensitive endpoints.

An additional aspect to consider include whether there are phylogenetic differences in physiology across species that can provide a target site for any pharmaceutical MOA, and/or sensitive life stages (e.g., developmental). Aquatic toxicity testing programs should attempt to protect (1) keystone species and biodiversity, (2) ecosystem structure and function, including primary productivity and important phyla, and (3) economically, commercially, and socially important species.

Similar to programs outlined by the EU (EMEA 1998) and USFDA (FDA-CDER 1998), Lange and Dietrich (2002) presented a basic strategy for developing an ecotoxicological testing scheme for xenobiotics that can also be applied pharmaceutical compounds. These authors suggest a two-tiered strategy, in which algae, *Daphnia*, and fish are examined in the first tier, using standardized testing procedures. Based on the results of first-tier testing, a second tier was included to test compounds that possess significant risk. Although this strategy is useful for many compounds, it lacks specificity for pharmaceuticals that have specific MOAs, and it does not incorporate sensitive or native species. Ankley et al. (2005) and Cunningham et al. (2006) partially address these shortcomings by suggesting that a standardized list of pharmaceuticals with specific MOAs be tested; such pharmaceuticals should represent different chemical classes to allow developing extensive ecotoxicological information that pertains to each MOA for the purpose of determining the potential risks of each general class. Ankley et al. (2007) further advanced this concept by providing a framework for evaluating sublethal and in vitro responses that would justify undertaking chronic studies; this framework was then integrated in a tiered-testing scheme proposed by Brooks et al. (2008).

As previously mentioned, our review of the acute toxicity posed by pharmaceutical compounds that reach the aquatic environment suggests that there is little concern for the actual environmental concentrations that occur. In developed countries, the acute effect concentrations are 100–1,000 times higher than the residue levels that are currently observed for many compounds. Therefore, acute toxicity and acute toxicity testing would occur only from spill events or other mass loading events, such as in developing countries that have limited regulatory or enforcement oversight (Carlsson et al. 2009). Although single chemical acute toxicity testing may produce very limited data, toxicity testing of mixtures, or of whole effluent toxicity (WET) could indicate effects not observed in single chemical acute testing. In the USA, WET testing refers to the aggregate toxic responses of aquatic organisms from the potential combined effects of all substances in a complex effluent (USEPA 1994; USEPA 2000; USEPA 2002a, b, c). WET testing results provide data for acute exposures on the combined effect of effluent and can be used to evaluate whether short- and/or long-term effects are possible (USEPA 1994). The same organism and endpoints (e.g., cladoceran mortality and reproduction) are employed for WET testing to develop water quality criteria and environmental standards. However, for an all-inclusive assessment of pharmaceutical effects, using WET testing and other toxicity studies for pharmaceuticals need not be limited to organisms presently employed (e.g., *C. dubia*, *P. promelas*), but rather should be augmented as necessary with additional model organisms and responses (e.g., fish endocrine function, behavior, reproduction).

Chronic toxicity studies examining pharmaceuticals have been increasing; however, data investigating key biological endpoint targets based on MOA is lacking (Berninger and Brooks 2010). Most chronic toxicity data have been developed from standardized methodology thus ignoring sensitive MOA-related endpoints. As stated above, MOA should guide the selection of endpoints for chronic toxicity testing and indicate potential effects due to exposure. Due to the broad range of pharmaceutical compounds, a priori knowledge of MOA is necessary for developing a toxicity testing plan (Daughton and Ternes 1999). WET testing can also be applied for chronic testing and can reveal important data (e.g., survival, growth, reproduction, development) on pharmaceutical mixtures. To date, only a minimal number of studies have examined chronic effects of pharmaceutical mixtures (e.g., Painter et al. 2009), especially in whole effluent scenarios, though mixture toxicity has been investigated for many other contaminants.

In the case of pharmaceuticals, selecting organisms and endpoints for study is possible because MOA data on target organisms are available, and the biological pathways targeted by pharmaceuticals are often conserved across species, particularly for vertebrates. Ankley et al. (2010) proposed adverse outcome pathways (AOP) as a conceptual framework to support ecological risk assessments of contaminants. The goal of the AOP is to create a stepwise linkage between molecular initiating events and the resulting adverse outcomes that occur at the organism and population levels (Fig. 2a). To illustrate the strength of this approach for ecological risk assessments of pharmaceuticals, an AOP for the pharmaceutical class of SSRIs was created (Fig. 2b). As noted above, SSRIs are designed to block SERTs, resulting in an increase of serotonin in presynaptic terminals. Comparable SSRI binding affinities for fish and rat SERTs and decreased SERT binding, which is related to therapeutic effects in mammals, has been observed in fish (Gould et al. 2007). At the organismal level, appetite suppression and behavioral modifications, including ecologically relevant feeding inhibition (Stanley et al. 2007; Gaworecki and Klaine 2008; Valenti et al. 2009) and predation avoidance (Painter et al. 2009), have been reported. When internal dose levels were consistent with mammalian C_{max} concentrations, sertraline significantly decreased SERT binding in the brain and resulted in significant behavioral effects in adult male fathead minnows (Valenti et al. 2012. See citation added to reference section).

In the future, we recommend that an AOP conceptual model be employed to guide and support selection of potentially sensitive model organisms and endpoints that are related to therapeutic MOAs, when laboratory, -cosm, or field assessments of pharmaceuticals are performed on WWTP effluents. We further recommend that

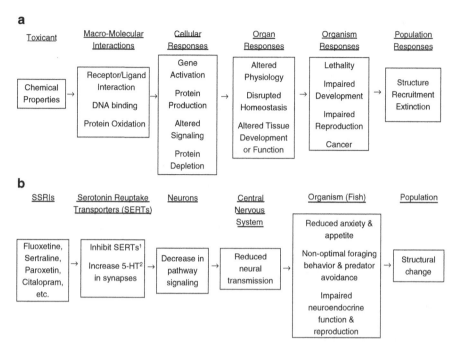

Fig. 2 Conceptual diagram of two adverse outcome pathways (AOPs), which highlight the linkages among molecular initiation events and anticipated responses at levels of increasing biological complexity. (**a**) The conceptual model of an AOP (modified from Ankley et al. 2010). (**b**) An example AOP for selective serotonin reuptake inhibitors (SSRIs). [1]Serotonin Reuptake Transporter. [2]5-Hydroxytryptamine

an abbreviated ecological risk assessment be conducted at the end of each tier with the application of weight of evidence (WOE) methods for risk characterization to determine possible impacts based on multiple lines of evidence (LOE; Burton et al. 2002; Chapman et al. 2002). The WOE approach should include at least four LOEs for an assessment of the effects of pharmaceuticals potentially present in the final effluent: (1) one based on contaminant characterization, including toxicity (and fate) data from the literature to estimate effects (and exposures); (2) one based on laboratory toxicity to surrogate; (3) one based on toxicity to native organisms; and (4) one concerning native biota community characterization (Long and Chapman 1985). Interactions with other stressors (e.g., nutrients) found in whole effluent discharges should be considered, particularly lower trophic levels exposed to antibiotics (Brooks et al. 2008; Fulton et al. 2009, 2010). When pharmaceuticals are ionizable (Nakamura et al. 2008; Valenti et al. 2009) or chiral (Stanley et al. 2006, 2007), the influence of site-specific pH (Valenti et al. 2011) and enantiomer-specific properties (Stanley and Brooks 2009) on aquatic toxicokinetics and toxicodynamics should be considered to reduce the uncertainty during ecological risk assessments.

7 Conclusions

Pharmaceuticals are continuously released into the environment from WWTP effluents. Residues of pharmaceutical compounds have been found in the aquatic environment throughout the world at low levels; however, because of continual release and replacement they are persistent compounds that could cause long-term ecological effects. In this review, we presented the growing array of endpoints and experiments that are associated with the environmental safety testing of pharmaceuticals that are currently being used or could be used to evaluate the effects of pharmaceuticals in WWTP effluent. As a result of performing this review of the pharmaceutical literature, we make the following conclusions:

1. As the population of the USA continues to increase, the number and amounts of pharmaceuticals discharged annually will continue to increase. Great strides have been made since Halling-Sorensen et al. (1998) and Daughton and Ternes (1999) first revealed the potential toxicological issues caused by PPCPs, but significant work is still required.
2. Sufficient research demonstrates that pharmaceuticals enter aquatic ecosystems and can produce adverse effects. Unfortunately, the scope and depth of research, thus far, is inadequate to determine the breadth of the effects caused.
3. There is currently a dearth of research data on many important aspects of pharmaceutical effects in aquatic ecosystems, including:

   (a) Chronic toxicity data for individual pharmaceuticals to benthic invertebrates, including bivalves, and fish is lacking;
   (b) Effects of pharmaceuticals on threatened or endangered species, which warrant protection at the individual level of biological organization;
   (c) MOA-based studies, in which biochemical and histological alterations are investigated or studies in which genetic alterations are monitored in response to long-term pharmaceutical exposure;
   (d) Techniques capable of detecting sensitive endpoints in aquatic organisms, such as in vitro and computational toxicology, for prioritizing chemicals and pathways for future studies;
   (e) Data on complex mixtures of pharmaceuticals that found in WWTP effluents;
   (f) Data on hazards to wildlife that depend on aquatic organisms for food (e.g., birds, reptiles, mammals) are needed, because pharmaceuticals are known to accumulate in aquatic life (Brooks et al. 2005; Chu and Metcalfe 2007; Ramirez et al. 2007, 2009; Brown et al. 2007; Zhou et al. 2008; Kwon et al. 2009; Fick et al. 2010; Schultz et al. 2010; Zhang et al. 2010; Daughton and Brooks 2011).

8 Summary

Although an increasingly large amount of data exists on the acute and chronic aquatic toxicity of pharmaceuticals, numerous questions still remain. There remains a dearth of information pertaining to the chronic toxicity of bivalves, benthic invertebrates, fish, and endangered species, as well as study designs that examine mechanism-of-action (MOA)-based toxicity, in vitro and computational toxicity, and pharmaceutical mixtures. Studies examining acute toxicity are prolific in the published literature; therefore, we address many of the shortcomings in the literature by proposing "intelligent" well-designed aquatic toxicology studies that consider comparative pharmacokinetics and pharmacodynamics. For example, few studies on the chronic responses of aquatic species to residues of pharmaceuticals have been performed, and very few on variables that are plausibly linked to any therapeutic MOA. Unfortunately, even less is understood about the metabolism of pharmaceuticals in aquatic organisms. Therefore, it is clear that toxicity testing at each tier of an ecological risk assessment scheme would be strengthened for some pharmaceuticals by selecting model organisms and endpoints to address ecologically problematic MOAs. We specifically recommend that future studies employ AOP approaches (Ankley et al. 2010) that leverage mammalian pharmacology information, including data on side effects and contraindications. Use of conceptual AOP models for pharmaceuticals can enhance future studies in ways that assist in the development of more definitive ecological risk assessments, identify chemical classes of concern, and help protect ecosystems that are affected by WWTP effluent discharge.

Acknowledgments We thank Jason P. Berninger for useful discussions and assistance with information used to develop Figure 1.

References

Andersen L, Kinnberg K, Holbech H, Korsgaard B, Bjerregaard P (2004) Evaluation of a 40 day assay for testing endocrine disruptors: effects of an anti-estrogen and aromatase inhibitor on sex ratio and vitellogenin concentrations in juvenile zebrafish (*Danio rerio*). Fish Physiol Biochem 30:257–266

Andreozzi R, Caprio V, Ciniglia C, De Champdore M, Lo Giudice R, Marotta R, Zuccato E (2004) Antibiotics in the environment: occurrence in Italian STPs, fate, and preliminary assessment on algal toxicity of amoxicillin. Environ Sci Technol 38:6832–6838

Ankley DT, Kahl MD, Jensen KM, Hornung MW, Korte JJ, Makynen EA, Leino RL (2002) Evaluation of the aromatase inhibitor fadrozole in a short-term reproduction assay with the fathead minnow (*Pimephales promelas*). Toxicol Sci 67:121–130

Ankley DT, Jensen KM, Makynen EA, Kahl MD, Korte JJ, Hornung MW, Henry TR, Denny JS, Leino RL, Wilson VS, Cardon MC, Hartig PC, Gray LE (2003) Effects of the androgenic growth promoter 17-beta-trenbolone on fecundity and reproductive endocrinology of the fathead minnow. Environ Toxicol Chem 22:1350–1360

Ankley GT, Black MC, Hutchinson TH, Iguchi T (2005) A framework for assessing the hazard of pharmaceutical materials to aquatic species. In: Williams RT (ed) Human pharmaceuticals: assessing the impacts on aquatic ecosystems. SETAC, Pensacola, FL, pp 183–237

Ankley GT, Daston GP, Degitz SJ, Denslow ND, Hoke RA, Kennedy SW, Miracle AL, Perkins EJ, Snape J, Tillitt DE, Tyler CR, Versteeg D (2006) Toxicogenomics in regulatory ecotoxicology. Environ Sci Technol 40:4055–4065

Ankley GT, Brooks BW, Huggett DB, Sumpter JP (2007) Repeating history: pharmaceuticals in the environment. Environ Sci Technol 41:8211–8217

Ankley GT, Bennett RS, Erickson RJ, Hoff DJ, Horning MW, Johnson RD, Mount DR, Nichols JW, Russom CL, Schmieder PK, Serrano JA, Tietge JE, Villeneuve DL (2010) Adverse outcome pathways: a conceptual framework to support ecotoxicology research and risk assessment. Environ Toxicol Chem 29:730–741

Anon (1993) Acute toxicity to bluegill (*Lepomis macrochirus*) of the test substance ketorolac tromethamine from Radian Corporation in a 96-hr static non-renewal test. AnaltiKEM Environmental Lab, Houston, USA (AnalytiKEM Test Number 01628)

Arcand-Hoy LD, Benson WH (2001) Toxic responses of the reproductive system. In: Schlenk D, Benson WH (eds) Target organ toxicity in marine and freshwater teleosts. Taylor & Francis, London

Auerbach SB, Lundberg JF, Hjorth S (1995) Differential inhibition of serotonin release by 5-HT and NA reuptake blockers after systemic administration. Neuropharmacology 34:89–96

Babich H, Borenfreund E (1987) In vitro cytotoxicity of organic pollutants to bluegill sunfish (BF-2) cells. Environ Res 42:229–237

Bach TJ, Lichtenthaler HK (1983) Inhibition by mevinolin of plant growth, sterol formation and pigment accumulation. Physiol Plant 59:50–60

Baldwin WS, Milam DL, Leblanc GA (1995) Physiological and biochemical perturbations in *Daphnia magna* following exposure to the model environmental estrogen diethylstilbestrol. Environ Toxicol Chem 14:945–952

Bantle JA, Burton DT, Dawson DA, Dumont JN, Finch RA, Fort DJ (1994) FETAX interlaboratory validation study: Phase II testing. Environ Toxicol Chem 13:1629–1637

Basset GJC, Quinlivam EP, Gregory JF, Hanson AD (2005) Folate synthesis and metabolism in plants and prospects for biofortification. Crop Sci 45:449–453

Belfroid A, Leonards P (1996) Effect of ethinyl oestradiol on the development of snails and amphibians. SETAC 17th Annual Meeting November 1996, Washington, DC, USA

Bell AM (2001) Effects of an endocrine disrupter on courtship and aggressive behavior of male three-spine stickleback, *Gasteroseus aculeatus*. Anim Behav 62:775–780

Berninger JP, Brooks BW (2010) Leveraging mammalian pharmaceutical toxicology and pharmacology data to predict chronic fish responses to pharmaceuticals. Toxicol Lett 193:69–78

Berninger JP, Du B, Connors KA, Eytcheson SA, Kolkmeier MA, Prosser KN, Valenti TW, Chambliss CK, Brooks BW (2011) Effects of the antihistamine diphenhydramine to select aquatic organisms. Environ Toxicol Chem 30:2065–2072

Bird SB, Gaspari RJ, Lee WJ, Dickson EW (2002) Diphenhydramine as a protective agent in rat model of acute, lethal organophosphate poisoning. Acad Emerg Med 9:1369–1372

Black MC, Connors DE, Rogers ED, Kwon J, Armbrust K (2009) Effects of fluoxetine and sertraline on growth and development of the African Clawed Frog (*Xenopus laevis*). In: Platform presentation at 30th annual Society of Environmental Toxicology and Chemistry. November 20–23

Brain RA, Cedergreen N (2008) Biomarkers in aquatic plants: selection and utility. Rev Environ Contam Toxicol 198:49–109

Brain RA, Johnson DJ, Richards SM, Sanderson H, Sibley PK, Solomon KR (2004a) Effects of 25 pharmaceutical compounds to *Lemna gibba* using a seven-day static-renewal test. Environ Toxicol Chem 23:371–382

Brain RA, Johnson DJ, Richards SM, Hanson ML, Sanderson H, Lam MW, Young C, Mabury SA, Sibley PK, Solomon KR (2004b) Microcosm evaluation of the effects of an eight pharmaceutical mixture to the aquatic macrophytes *Lemna gibba* and *Myriophyllum sibricum*. Aquat Toxicol 70:23–40

Brain RA, Reitsma TS, Lissemore LI, Bestari B-J, Sibley PK, Solomon KR (2006a) Herbicidal effects of statin pharmaceuticals in *Lemna gibba*. Environ Sci Technol 40:5116–5123

Brain RA, Sanderson H, Sibley PK, Solomon KR (2006b) Probabilistic ecological hazard assessment: evaluating pharmaceutical effects on aquatic higher plants as an example. Ecotoxicol Environ Saf 6:128–135

Brain RA, Ramirez AJ, Fulton BA, Chambliss CK, Brooks BW (2008a) Herbicidal effects of sulfamethoxazole in *Lemna gibba*: Using p-aminobenzoic acid as a biomarker of effect. Environ Sci Technol 42:8965–8970

Brain RA, Hanson ML, Solomon KR, Brooks BW (2008b) Aquatic plants exposed to pharmaceuticals: effects and risks. Rev Environ Contam Toxicol 192:67–115

Brambilla G, Civitareale C, Migliore L (1994) Experimental toxicity and analysis of bacitracin, flumequine and sulphadimethoxine in terrestrial and aquatic organisms as a predictive model for ecosystem damage. Qumica Analitica 13(Suppl):S73–S77

Breitholtz M, Bengtsson BE (2001) Oestrogens have no hormonal effect on the development and reproduction of the harpacticoid copepod *Nitocra spinepes*. Mar Pollut Bull 42:879–886

Breton R, Boxall A (2003) Pharmaceuticals and personal care products in the environment: regulatory drivers and research needs. QSAR Comb Sci 22:399–409

Bringmann G, Kuhn R (1982) Ergebnisse der Schadwirkung wassergefahrdender Stoffe gegen *Daphnia magna* in einem weiterentwickelten standardisierten Testverfahren. Z Wasser Abwasser Forsch 15(1):1–6

Bringolf RB, Heltsley RM, Newton TJ, Eads CB, Fraley SJ, Shea D, Cope WG (2010) Environmental occurrence and reproductive effects of the pharmaceutical fluoxetine in native freshwater mussels. Environ Toxicol Chem 29:1311–1318

Brooks BW, Foran CM, Richards SM, Weston J, Turner PK, Stanley JK, Solomon KR, Slattery M, La Point TW (2003a) Aquatic ecotoxicology of fluoxetine. Toxicol Lett 142:169–183

Brooks BW, Turner PK, Stanley JK, Weston JJ, Glidewell EA, Foran CM, Slattery M, La Point TW, Huggett DB (2003b) Waterborne and sediment toxicity of fluoxetine to select organisms. Chemosphere 52:135–142

Brooks BW, Richards SM, Weston J, Turner PK, Stanley JK, La Point TW, Brain R, Glidewell EA, Massengale ARD, Smith W, Blank CL, Solomon KR, Slattery M, Foran CM (2005) Aquatic ecotoxicology of fluoxetine: a review of recent research. In: Dietrich D, Webb S, Petry T (eds) Hot spot pollutants: pharmaceuticals in the environment. Academic/Elsevier, London, UK, pp 165–187

Brooks BW, Riley TM, Taylor RD (2006) Water quality of effluent-dominated stream ecosystems: ecotoxicological, hydrological, and management considerations. Hydrobiologia 556:365–379

Brooks BW, Ankley GT, Hobson JF, Lazorchak JM, Meyerhoff RD, Solomon KR (2008) Assessing the aquatic hazards of veterinary medicines. In: Crane M, Boxall A, Barrett K (Eds) Veterinary medicines in the environment. CRC/SETAC Press. pp 97–128

Brooks BW, Huggett DB, Boxall ABA (2009a) Pharmaceuticals and personal care products: research needs for the next decade. Environ Toxicol Chem 28:2469–2472

Brooks BW, Huggett DB, Brain RA, Ankley GT (2009b) Risk assessment considerations for veterinary medicines in aquatic systems. In: Henderson K, Coats J (eds) Veterinary pharmaceuticals in the environment. American Chemical Society, Washington DC, pp 205–223

Brown JN, Paxeus N, Forlin L, Larsson DGJ (2007) Variations in bioconcentration of human pharmaceuticals from sewage effluents into fish blood plasma. Environ Toxicol Pharmacol 24:267–274

Bruschweiler BJ, Wurgler FE, Fent K (1995) Cytotoxicity in vitro of organotin compounds to fish hepatoma cells PLHC-1 (*Poeciliopsis lucida*). Aquat Toxicol 32:143–160

Burton GA, Chapman PM, Smith EP (2002) Weight-of-evidence approaches for assessing ecosystem impairment. Human Ecol Risk Assess 8:1657–1673

Buttgereit F, Burmester GR, Simon LS (2001) Gastrointestinal toxic side effects of nonsteroidal anti-inflammatory drugs and cyclooxygenase-2-specific inhibitors. Am J Med 110:13–19

Calleja MC, Personne G, Geladi P (1994) Comparative acute toxicity of the first 50 multicentre evaluation of in vitro cytotoxicity chemicals to aquatic non-vertebrates. Arch Environ Contam Toxicol 26:69–78

Caminada D, Escher C, Fent K (2006) Cytotoxicity of pharmaceuticals found in aquatic systems: comparison of PLHC-1 and RTG-2 fish cell lines. Aquat Toxicol 79:114–123

Carlsson G, Orn S, Larsson DGJ (2009) Effluent from bulk drug production is toxic to aquatic vertebrates. Environ Toxicol Chem 28:2656–2662

Chapman PM, McDonald BG, Lawrence GS (2002) Weight-of-evidence issues and frameworks for sediment quality (and other) assessments. Human Ecol Risk Assess 8:1489–1515

Christensen AM, Faaborg-Andersen S, Ingerslev F, Baun A (2007) Mixture and single-substance toxicity of selective serotonin reuptake inhibitors toward algae and crustaceans. Environ Toxicol Chem 26:85–91

Chu S, Metcalfe CD (2007) Analysis of paroxetine, fluoxetine and norfluoxetine in fish tissues using pressurized liquid extraction, mixed mode solid phase extraction cleanup and liquid chromatography-tandem mass spectrometry. J Chromatogr A 1163:112–118

Clements S, Schreck CB (2007) Chronic administration of fluoxetine alter locomotor behavior, but does not potentiate the locomoter stimulating effects of CRH in juvenile Chinook salmon (*Onchorhynchus tshawytscha*). Comp Biochem Physiol A Mol Integr Physiol 147(1):43–49

Cleuvers M (2003) Aquatic ecotoxicity of pharmaceuticals including the assessment of combination effects. Toxicol Lett 142:185–194

Cleuvers M (2004) Mixture toxicity of the anti-inflammatory drugs diclofenac, ibuprofen, naproxen, and acetylsalicylic acid. Ecotoxicol Environ Saf 59:309–315

Cleuvers M (2005) Initial risk assessment for three β-blockers found in the aquatic environment. Chemosphere 59:199–205

Clotfelter ED, O'Hare EP, McNitt MM, Carpenter RE, Summers CH (2007) Serotonin decreases aggression via 5-HT1A receptors in the fighting fish *Betta splendens*. Pharmacol Biochem Behav 87(2):222–231

Clubbs RL, Brooks BW (2007) Daphnia magna responses to a vertebrate estrogen receptor agonist and an antagonist: a multigenerational study. Ecotoxicol Environ Saf 67:385–398

Coats JR, Metcalf RL, Lu P-Y, Brown DO, Williams JF, Hansen LG (1976) Model ecosystem evaluation of the environmental impacts of the veterinary drugs phenothiazine, sulfametazine, clopidol and diethylstibestrol. Environ Health Perspect 1:167–197

Collinge J, Pycock CJ, Taberner PV (1983) Studies on the interaction between cerebral 5-hydroxytryptamine and gamma-aminobutyric acid in the mode of action of diazepam in the rat. Br J Pharmacol 79:637–643

Conners DE, Rogers ED, Armbrust KL, Kwon J-W, Black MC (2009) Growth and development of tadpoles (*Xenopus laevis*) exposed to selective serotonin reuptake inhibitors, fluoxetine and sertraline, throughout metamorphosis. Environ Toxicol Chem 28:2671–2676

Corcoran J, Winter MJ, Tyler CR (2010) Pharmaceuticals in the aquatic environment: a critical review of the evidence for health effects in fish. Crit Rev Toxicol 40:287–304

Crane M, Watts C, Boucard T (2006) Chronic aquatic environmental risks from exposure to human pharmaceuticals. Sci Total Environ 367:23–41

Crespi EJ, Denver RJ (2004) Ontogeny of corticotrophin-releasing factor effects on locomotion and foraging in the Western spadefoot toad (*Spea hammondii*). Horm Behav 46:399–410

Cunningham VL, Constable DJC, Hannah RE (2004) Environmental risk assessment of paroxetine. Environ Sci Technol 38:3351–3359

Cunningham VL, Buzby M, Hutchinson T, Mastrocco F, Parke N, Roden N (2006) Effects of human pharmaceuticals on aquatic life: next steps. Environ Sci Technol 40:3457–3462

Czech P, Weber K, Dietrich DR (2001) Effects of endocrine modulating substances on reproduction in the hermaphroditic snail *Lymnaea stagnalis* L. Aquat Toxicol 53:103–114

Dahl U, Gorokhova E, Breitholtz M (2006) Application of growth-related sublethal endpoints in ecotoxicological assessments using a harpacticoid copepod. Aquat Toxicol 77:433–438

Daston GP, Gooch JW, Breslin WJ, Shuey DL, Nikiforov AI, Ficao TA, Gorsuch JW (1997) Environmental estorgens and reproductive health: a discussion of the human and environmental data. Reprod Toxicol 11:465–481

Daughton CG (2002) Environmental stewardship and drugs as pollutants. Lancet 360:1035–1036

Daughton CG, Brooks BW (2011) Active pharmaceuticals ingredients and aquatic organisms. In: Environmental contaminants in wildlife: interpreting tissue concentrations, 2nd edn. Meador J, Beyer N (eds) Taylor and Francis, pp 281–341

Daughton CG, Ternes TA (1999) Pharmaceuticals and personal care products in the environment: agents of subtle change? Environ Health Perspect 107:907–938

David A, Pancharata K (2009) Developmental anomalies induced by a non-selective COX inhibitor (ibuprofen) in zebrafish (*Danio rerio*). Environ Toxicol Pharmacol 27:390–395

De Lange HJ, Noordoven W, Murk AJ, Lurling M, Peeters ETHM (2006) Behavioural responses of *Gammarus pulex* (Crustacea, Amphipoda) to low concentrations of pharmaceuticals. Aquat Toxicol 78:209–216

De Liguoro M, Fioreto B, Poltronieri C, Gallina G (2009) The toxicity of sulfamethazine to *Daphnia magna* and its additivity to other veterinary sulfonamides and trimethoprim. Chemosphere 75:1519–1524

De Young DJ, Bantle JA, Hull MA, Burks SL (1996) Differences in the sensitivity to developmental toxicants as seen in *Xenopus* and *Pimephales* embryos. Bull Environ Contam Toxicol 56:143–150

DellaGreca M, Iesce MR, Isidori M, Nardelli A, Previtera L, Rubino M (2007) Phototransformation products of tamoxifen by sunlight in water. Toxicity of the drug and its derivatives on aquatic organisms. Chemosphere 67:1933–1939

DeLorenzo ME, Fleming J (2008) Individual and mixture effects of selected pharmaceuticals and personal care products on the marine phytoplankton species *Dunaliella tertiolecta*. Arch Environ Contam Toxicol 54:203–210

Denver RJ (1997) Proximate mechanisms of phenotypic plasticity in amphibian metamorphosis. Am Zool 37:172–184

Di Delupis GD, Macri A, Civitareale C, Migliore L (1992) Antibiotics of zootechnical use: effects of acute high and low dose contamination on *Daphnia magna* Straus. Aquat Toxicol 22:53–60

Dinchuk JE, Car BD, Focht RJ, Johnston JJ, Jaffee BD, Covington MB, Contel NR, Eng VM, Collins RJ, Czerniak PM, Gorry SA, Trzaskos JM (1995) Renal abnormalities and an altered inflammatory response in mice lacking cyclooxygenase II. Nature 378:406–409

Disch A, Hemmerlin A, Bach TJ, Rohmer M (1998) Mevalonate-derived isopentyl diphosphate is the biosynthetic precursor of ubiquinone prenyl side chain in tobacco BY-2 cells. Biochem J 331:615–621

Dobbins LL, Brain RA, Brooks BW (2008) Comparison of the sensitivities of common in vitro and in vivo assays of estrogenic activity: application of chemical toxicity distributions. Environ Toxicol Chem 27:2608–2616

Dobbins LL, Usenko S, Brain RA, Brooks BW (2009) Probabilistic ecological hazard assessment of parabens using *Daphnia magna* and Pimephales promelas. Environ Toxicol Chem 28:2744–2753

Dorne JLCM, Ragas AMJ, Frampton GK, Spurgeon DS, Lewis DF (2007) Trends in human risk assessment of pharmaceuticals. Anal Bioanal Chem 387:1167–1172

Dussault EB, Balakrishnan VK, Sverko E, Solomon KR, Sibley PK (2008) Toxicity of human pharmaceuticals and personal care products to benthic invertebrates. Environ Toxicol Chem 27:425–432

Dzialowski EM, Turner PK, Brooks BW (2006) Physiological and reproductive effects of beta adrenergic receptor antagonists in *Daphnia magna*. Arch Environ Contam Toxicol 50:503–510

Eguchi K, Nagase H, Ozawa M, Endoh YS, Goto K, Hirata K, Miyamoto K, Yoshimura H (2004) Evaluation of antimicrobial agents for veterinary use in the ecotoxicity test using microalgae. Chemosphere 57:1733–1738

Elvers KT, Wright SJL (1995) Antibacterial activity of the anti-inflammatory compounds ibuprofen. Lett Appl Microbiol 20:82–84

Emblidge JP, DeLorenzo ME (2006) Preliminary risk assessment of the lipid-regulating pharmaceutical clofibric acid, for three estuarine species. Environ Res 100:216–226

EMEA (1998) Note for guidance: environmental risk assessment of veterinary medicinal products other than GMO-containing and immunological products. EMEA, London (EMEA/CVMP/055/96)

EMEA (2005) Note for guidance on environmental risk assessment of medicinal products for human use, CMPC/SWP/4447/draft. The European Agency for the Evaluation of Medicinal Products (EMEA), London

EMEA (2006) Guideline on the environmental risk assessment of medicinal products for human use, EMEA/CHMP/4447/00. The European Agency for evaluation of medicinal products, London

Eravci M, Pinna G, Meinhold H, Baumgartner A (2000) Effects of pharmacological and nonpharmacological treatments of thyroid hormone metabolism and concentrations in rat brain. Endocrinology 141:1027–1040

Escher BI, Hermens JLM (2002) Modes of action in ecotoxicology: their role in body burdens, species sensitivity, QSARs, and mixture effects. Environ Sci Technol 36:4201–4217

Escher BI, Bramaz N, Eggen RIL, Richter M (2005) In vitro assessment of modes of toxic action of pharmaceuticals in aquatic life. Environ Sci Technol 39:3090–3100

European Union (2001) White Paper; Strategy for a future chemicals policy. COM 2001. Brussels

Faxon DP, Halperin JL, Creager MA, Gavras H, Schick EC, Ryan TJ (1981) Angiotensin inhibition in severe heart failure: acute central and limb hemodynamic effects of captopril with observations on sustained oral therapy. Am Heart J 101:548–556

FDA (1996) Retrospective review of ecotoxicity data submitted in environmental assessments. US Food and Drug Administration, Center for Drug Evaluation and Research (CDER), Rockville, MD

FDA (1998) Evidence for industry for the submission of an environmental in human drug applications and supplements. US Food and Drug Administration, Center for Drug Evaluation and Research (CDER) and Center for Biologics Evaluation and Research, Rockville, MD

Fent K, Weston AA, Caminada D (2006) Ecotoxicology of human pharmaceuticals. Aquat Toxicol 76:122–159

Ferrari B, Paxeus N, Lo Giudice R, Pollio A, Garric J (2003) Ecotoxicological impact of pharmaceuticals found in treated wastewaters: study of carbamazepine, clofibric acid, and diclofenac. Ecotoxicol Environ Saf 55:359–370

Ferrari B, Mons R, Vollat B, Frayse B, Paxeus N, Lo Guidice R, Pollio A, Garric J (2004) Environmental risk assessment of six pharmaceuticals: are the current environmental risk assessment procedures sufficient for the protection of the aquatic environment? Environ Toxicol Chem 23:1344–1354

Fick J, Lindberg RH, Parkkonen J, Arvidsson B, Tysklind M, Larsson DGJ (2010) Therapeutic levels of levonorgestrel detected in blood plasma of fish: results from screening rainbow trout exposed to treated sewage effluents. Environ Sci Technol 44:2661–2666

Flippin JL, Huggett D, Foran CM (2007) Changes in the timing of reproduction following chronic exposure to ibuprofen in Japanese medaka (Oryzias latipes). Aquat Toxicol 81:73–83

Fong PP, Huminski PT, D'Urso LM (1998) Induction and potentiation of parturition in fingernail clams (Sphaerium striatinum) by selective serotonin re-uptake inhibitors (SSRIs). J Exp Zool 280:260–264

Foran CM, Weston J, Slattery M, Brooks BW, Huggett DB (2004) Reproductive assessment of Japanese medaka (Oryzias latipes) following a four-week fluoxetine (SSRI) exposure. Arch Environ Contam Toxicol 46:511–517

Foster HR, Burton GA, Basu N, Werner EE (2010) Chronic exposure to fluoxetine (Prozac) causes developmental delays in Rana pipiens larvae. Environ Toxicol Chem 29(12):2845–2850

Fraker SL, Smith GR (2004) Direct and indirect effects of ecologically relevant concentrations of organic wastewater contaminants on Rana pipiens tadpoles. Environ Toxicol 19:250–256

Fujioka Y (2002) Effetcs of hormone treatments and temperature on sex-reversal of Nigorobuna Carassius carassius grandoculis. Fish Sci 68:889–893

Fulton BA, Brain RA, Usenko S, Back JA, King RS, Brooks BW (2009) Influence of N and P concentrations and ratios on Lemna gibba growth responses to triclosan in laboratory and stream mesocosm experiments. Environ Toxicol Chem 28:2610–2621

Fulton BA, Brain RA, Usenko S, Back JA, Brooks BW (2010) Exploring Lemna gibba thresholds to nutrient and chemical stressors: differential effects of triclosan on internal stoichiometry and nitrate uptake across a N:P gradient. Environ Toxicol Chem 29:2363–2370

Furr BJA, Tucker H (1996) The preclinical development of bicalutamide: pharmacodynamics and mechanism of action. Urology 47:13–25

Gaworecki KM, Klaine SJ (2008) Behavioral and biochemical responses of hybrid striped bass during and after fluoxetine exposure. Aquat Toxicol 88:207–213

Gledhill WE, Feijtel TCJ (1992) Environmental properties and safety assessment of organic phosphonates used for detergent and water treatment. In: de Oude NT (ed) Detergents. Handbook of environmental chemistry. Springer, New York, pp 261–285

Gould GG, Brooks BW, Frazer A (2007) [³H] Citalopram binding to serotonin transporter sites in minnow brains. Basic Clin Pharmacol Toxicol 101:203–210

Gravel A, Wilson JM, Pedro DFN, Vijayan MM (2009) Non-steroidal anti-inflammatory drugs disturb the osmoregulatory, metabolic and cortisol responses associated with seawater exposure in rainbow trout. Comp Biochem Physiol C 149:481–490

Grisolia CK (2002) A comparison between mouse and fish micronucleus test using cyclophosphamide, mitomycin C and various pesticides. Mutat Res 518:145–150

Grung M, Kallqvist T, Sakshaug S, Skurtveit S, Thomas KV (2008) Environmental assessment of Norwegian priority pharmaceuticals based on the EMEA guideline. Ecotoxicol Environ Saf 71:328–340

Guler Y, Ford AT (2010) Anti-depressants make amphipods see the light. Aquat Toxicol 99:397–404

Gyllenhammar I, Holm L, Eklund R, Berg C (2009) Reproductive toxicity in *Xenopus tropicalis* after developmental exposure to environmental concentrations of ethynylestradiol. Aquat Toxicol 91:171–178

Haider S, Baqri SSR (2000) β-Adrenoceptor antagonists reinitiate meiotic maturation in *Clarias batrachu* oocytes. Comp Biochem Physiol A 126:517–525

Halling-Sorensen B (2000) Algal toxicity of antibacterial agents used in intensive farming. Chemosphere 40:731–739

Halling-Sorensen B, Nielsen SN, Lanzky PF, Ingerster F, Lutzhoff HCH, Jorgensen SE (1998) Occurrence, fate and effects of pharmaceutical substances in the environment-a review. Chemosphere 36:357–393

Han GH, Hur HG, Kim SD (2006) Ecotoxicological risk of pharmaceuticals from wastewater treatment plants in Korea: occurrence and toxicity to *Daphnia magna*. Environ Toxicol Chem 25:265–271

Harrass MC, Kindig AC, Taub FB (1985) Responses of blue-green and green algae to streptomycin in unialgal and paired culture. Aquat Toxicol 6:1–11

Hasselberg L, Grosvik B-E, Goksoyr A, Celander MC (2005) Interactions between xenoestrogens and ketoconazole on hepatic CYP1A and CYP3A in juvenile Atlantic cod (*Gadhus mortiua*). Comp Hepatol 4:2

Heath RJ, Rubin JR, Holland DR, Zhang E, Snow ME, Rock CO (1999) Mechanisms of triclosan inhibition of bacterial fatty acid synthesis. J Biol Chem 274:11110–11114

Heckmann LH, Callaghan A, Hooper HL, Connon R, Hutchinson TH, Mound SJ, Sibley RM (2007) Chronic toxicity of ibuprofen to *Daphnia magna*: effects on life history traits and population dynamics. Toxicol Lett 172:137–145

Heddle JA, Cimino MC, Hayashi M, Romagna MD, Shelby JD, Vanparys P, MacGregor JT (1991) Micronuclei as an index of cytogenetic damage: past, present, and future. Environ Mol Mutagen 18:277–291

Hegelund T, Ottoson K, Radinger M, Tomberg P, Celander MC (2004) Effects of the antifungal imidazole ketoconazole on CYP1A and CYP3A in rainbow trout and killifish. Environ Toxicol Chem 23:1326–1334

Henry TB, Black MC (2008) Acute and chronic toxicity of fluoxetine (selective serotonin reuptake inhibitor) in western mosquito fish. Arch Environ Contam Toxicol 54:325–330

Henry TB, Kwon JW, Armbrust KL, Black MC (2004) Acute and chronic toxicity of five selective serotonin reuptake inhibitors in *Ceriodaphnia dubia*. Environ Toxicol Chem 23:2229–2233

Henschel KP, Wenzel A, Diederich M, Fliedner A (1997) Environmental hazard assessment of pharmaceuticals. Regul Toxicol Pharmacol 25:220–225

Hernando MD, Mezcua M, Fernandez-Alba AR, Barcelo D (2006) Environmental risk assessment of pharmaceutical residues in wastewater effluents, surface waters and sediments. Talanta 69:334–342

Higniite C, Azarnoff DL (1977) Drugs and drug metabolites as environmental contaminants: chlorophenoxyisobutyrate and salicylic acid in sewage water effluent. Life Sci 20:337–341

Hinz B, Cheremina O, Brune K (2008) Acetaminophen (paracetamol) is a selective cyclooxygenase-2 inhibitor in man. FASEB J 22:383–390

Hoeger B, Kollner B, Dietrich DR, Hitzfeld B (2005) Water-borne diclofenac affects kidney and gill integrity and selected immune parameters in brown trout (*Salmo trutta* f. *fario*). Aquat Toxicol 75:53–64

Hogan NS, Duarte P, Wade MG, Lean DRS, Trudeau VL (2008) Estrogenic exposure affects metamorphosis and alters sex ratios in the northern leopard frog (*Rana pipiens*): identifying critically vulnerable periods of development. Gen Comp Endocrinol 156:515–523

Holten Lutzhoft HC, Halling-Sorensen B, Jorgensen SE (1999) Algal testing of antibiotics applied in Danish fish farming. Arch Environ Contam Toxicol 36:1–6

Huet M-C (2000) OECD activity on endocrine disrupters test guidelines development. Ecotoxicology 9:77–94

Huggett DB, Brooks BW, Peterson B, Foran CM, Schlenk D (2002) Toxicity of select beta adrenergic receptor-blocking pharmaceuticals (B-blockers) on aquatic organisms. Arch Environ Contam Toxicol 43:229–235

Huggett DB, Cook JC, Ericson JF, Williams RT (2003) A theoretical model for utilizing mammalian pharmacology and safety data to prioritize potential impacts of human pharmaceuticals to fish. Hum Ecol Risk Assess 9:1789–1799

Hughes JS (1973) Acute toxicity of thirty chemicals to striped bass *(Marone saxatilis)*. In: Presented at the Western Association of State Game and Fish Commissioners in Salt Lake City, Utah

Hutchinson TH (2002) Reproductive and developmental effects of endocrine disruptors in invertebrates: in vitro and in vivo approaches. Toxicol Lett 131:75–81

Hutchinson TH, Yokota H, Hagino S, Ozato K (2003) Development of fish tests for endocrine disruptors. Pure Appl Chem 75:2343–2353

Hutchinson TH, Ankley GT, Segner H, Tyler CR (2006) Screening and testing for endocrine disruption in fish – Biomarker as "signposts", not "traffic lights", in risk assessment. Environ Health Perspect 114:106–114

Isidori M, Lavorgna M, Nardelli A, Parrella A, Previtera L, Rubino M (2005a) Ecotoxicity of naproxen and its phototransformation products. Sci Total Environ 348:93–101

Isidori M, Lavorgna M, Nardelli A, Pascarelli L, Parrella A (2005b) Toxic and genotoxic evaluation of six antibiotics on non-target organisms. Sci Total Environ 346:87–98

Isidori M, Nardelli A, Pascarella L, Rubino M, Parrella A (2007) Toxic and genotoxic impact of fibrates and their photoproducts on non-target organism. Environ Int 33:635–641

Jobling S, Sheahan D, Osborne JA, Matthiessen P, Sumpter JP (1996) Inhibition of testicular growth in rainbow trout (*Oncorhynchus mykiss*) exposed to estrogenic alkylphenolic chemicals. Environ Toxicol Chem 15:194–202

Johnson SK (1976) Twenty-four hour toxicity tests of six chemicals to mysis larvae of *Penaeus setiferus*. Texas A & M University Extension, Disease Laboratory (Publication No. FDDL-S8)

Johnson DJ, Sanderson H, Brain RA, Wilson CJ, Solomon KR (2007) Toxicity and hazard of selective serotonin reuptake inhibitor antidepressants fluoxetine, fluvoxamine, and sertraline to algae. Ecotoxicol Environ Saf 67:128–139

Jones OA, Voulvoulis N, Lester JN (2002) Aquatic environmental assessment of the top 25 English prescription pharmaceuticals. Water Res 36:5013–5022

Jørgensen SE, Halling-Sørensen B (2000) Drugs in the environment. Chemosphere 40:691–699

Jos A, Repetto G, Rios JC, Hazen N, Molero ML, del Peso A, Salguero M, Fernandez-Freire P, Perez-Martin JM, Camen A (2003) Ecotoxicological evaluation of carbamezapine using six different model systems with eighteen endpoints. Toxicol In Vitro 17:525–532

Kang IJ, Yokota H, Oshima Y, Tsuruda Y, Yamaguchi T, Maeda M, Imada N, Tadokoro H, Honjo T (2002) Effects of 17β-estradiol on the reproduction of Japanese medaka (*Oryzias latipes*). Chemosphere 47:71–80

Key PB, Hoguet J, Reed LA, Chung KW, Fulton MH (2008) Effects of the statin antihyperlipidemic agent simvastatin on grass shrimp, *Palaemonetes pugio*. Environ Toxicol 23:153–160

Kidd KA, Blanchfield PJ, Mills KH, Palance VP, Evans RE, Lazorchak JM, Flick RM (2007) Collapse of a fish population after exposure to a synthetic estrogen. Proc Natl Acad Sci USA 104:8897–8901

Kim Y, Choi K, Jung J, Park S, Kim P-G, Park J (2007) Aquatic toxicity of acetaminophen, carbamazepine, cimetidine, diltiazem and six major sulfonamides, and their potential ecological risks in Korea. Eniviron Int 33:370–375

Kim J-W, Ishibashi H, Yamauchi R, Ichikawa N, Takao Y, Hirano M, Koga M, Arizono K (2009) Acute toxicity of pharmaceutical and personal care products on freshwater crustacean (*Thamnocephalus platyurus*) and fish *(Oryzias latipes)*. J Toxicol Sci 34:227–232

Kim J, Park J, Kin PG, Lee C, Choi K, Choi K (2010) Implications of global environmental changes on chemical toxicity-effect of water temperature, pH, and ultraviolet B irradiation on acute toxicity of several pharmaceuticals in *Daphnia magna*. Ecotoxicology 19:662–669

Kirkegaard C, Faber J (1998) The role of thyroid hormones in depression. Eur J Endocrinol 138:1–9

Kirpichnikov D, McFarlane S, Sowers JR (2002) Metformin: an update. Ann Intern Med 137:25–33

Knoll/BASF (1995) Pharmaceutical safety data sheet ibuprofen (Issue/Revision 06/04/94). Knoll Pharmaceuticals, Nottingham, UK

Kollros JJ (1961) Mechanisms of amphibian metamorphosis: hormones. Am Zool 1:107–114

Kopf W (1995) Wirkung endokriner Stoffe in Biotest mit Wasserorganismen. Vortag bei der 50. Fachtagung des Bay. LA fur Wasserwirtschaft: Stoffe mit endokriner. Wirkung im Wasser

Kreke N, Dietrich DR (2008) Physiological endpoints for potential SSRI interactions in fish. Crit Rev Toxicol 37:215–247

Kuhn R, Pattard M, Pernak KD, Winter A (1989) Results of the harmful effects of selected water pollutants (anilines, phenols, aliphatic compounds) to *Daphnia magna*. Wat Res 23:495–499

Küster A, Alder AC, Escher BI, Duis K, Fenner K, Garric J, Hutchinson TH, Lapen DR, Pery A, Rombke J, Snape J, Ternes T, Topp E, Wehrhan A, Knacker T (2010) Environmental risk assessment of human pharmaceuticals in the European Union: a case study with the β-blocker atenolol. Integr Environ Assess Manag 6(Suppl 1):514–523

Kwon JW, Armbrust KL, Vidal-Dorsch D, Bay SM (2009) Determination of 17α-ethynylestradiol, carbamazepine, diazepam, simvastatin, and oxybenzone in fish livers. J AOAC Int 92:359–369

Laird BD, Brain RA, Johnson DJ, Wilson CJ, Sanderson H, Solomon KR (2007) Toxicity and hazard of a mixture of SSRIs to zooplankton communities evaluated in aquatic microcosms. Chemosphere 69(6):949–954

Lange R, Dietrich D (2002) Environmental risk assessment of pharmaceutical drug substances – conceptual considerations. Toxicol Lett 131:97–104

Lange R, Hutchinson TH, Croudace CP, Siegmund F, Schweinfurth H, Hampe P, Panter GH, Sumpter JP (2001) Effects of the synthetic estrogen 17 alpha-ethinylestradiol on the life cycle of the fathead minnow (*Pimephales promelas*). Environ Toxicol Chem 20:1216–1227

Lange IG, Daxenberger A, Schiffer B, Witters H, Ibarreta D, Meyer HHD (2002) Sex hormones originating from different livestock production systems: fate and potential endocrine disrupting activity in the environment. Anal Chim Acta 473:27–37

Lanzky PF, Halling-Sorenson B (1997) The toxic effects of the antibiotic metronidazole on aquatic organisms. Chemosphere 35:2553–2561

Larkin P, Folmar LC, Hemmer MJ, Poston A, Denslow ND (2003) Expression profiling of estrogenic compounds using a sheepshead minnow cDNA macroarray. Environ Health Perspect 111:839–846

Larsson DGJ, Hyllner SJ, Haux C (1994) Induction of vittelline envelope proteins by estradiol-17β in 10 teleost species. Gen Comp Endocrinol 96:445–450

Larsson DGJ, Adolfsson-Erici M, Parkkonen J, Pettersson M, Berg AH, Olsson P-E, Forlin L (1999) Ethinylestradiol-an undesired fish contraceptive? Aquat Toxicol 45:91–97

Laufs U, Liao JK (1998) Post-transcriptional regulation of endothelial nitric oxide synthase mRNA stability by RhoGTPase. J Biol Chem 273:24266–24271

Laville N, Ait-Aissa S, Gomez E, Casellas M, Porcher JM (2004) Effects of human pharmaceuticals on cytotoxicity, EROD activity and ROS production in fish hepatocytes. Toxicology 196:41–55

Lawrence JR, Swehone GDW, Wassenaar CI, Neu TF (2005) Effects of selected pharmaceuticals on riverine biofilm communities. Can J Microbiol 51:655–669

LeBlanc GA, Campbell PM, den Besten P, Brown RP, Chang ES, Coats JR, deFur PL, Dhadialla T, Edwards J, Riddiford LM, Simpson MG, Snell TW, Thorndyke M, Matsumura F (1999) The endocrinology of invertebrates. In: deFur PL, Crane M, Ingersoll CG, Tattersfield L (eds) Endocrine disruption in invertebrates: endocrinology, testing, and assessment. SETAC Press, Pensacola, FL

Li Z, Li P, Randak T (2010) Ecotoxicological effects of short-term exposure to a human pharmaceutical verapamil in juvenile rainbow trout (*Oncorhynchus mykiss*). Comp Biochem Physiol C 152:385–391

Li Z, Zlabek V, Velisek J, Grabic R, Machova J, Kolarova J, Li P, Randak T (2011) Acute toxicity of carbamazepine to juvenile rainbow trout (*Oncorhynchus mykiss*): effects on antioxidant responses, hematological parameters and hepatic EROD. Ecotoxicol Environ Saf 74:319–327

Lichtenthaler HK, Schwender J, Disch A, Rohmer M (1997a) Biosynthesis of isoprenoids in higher plant chloroplasts proceeds via a mevalonate-independent pathway. FEBS Lett 400:271–274

Lichtenthaler HK, Rohmer M, Schwender J (1997b) Two independent biochemical pathways for isopentyl diphosphate and isoprenoid biosynthesis in higher plants. Physiol Plant 101:643–652

Liebig M, Fernandez AA, Blübaum-Gronau E, Boxall A, Brinke M, Carbonell G, Egeler P, Fenner K, Fernandez C, Fink G, Garric J, Halling-Sørensen B, Knacker T, Krogh KA, Küster A, Löffler D, Cots MA, Pope L, Prasse C, Römbke J, Rönnefahrt I, Schneider MK, Schweitzer N, Tarazona JV, Ternes TA, Traunspurger W, Wehrhan A, Duis K (2010) Environmental risk assessment of ivermectin: a case study. Integr Environ Assess Manag 6:567–587

Lilius H, Isomaa B, Holmstrom T (1994) A comparison of the toxicity of 50 reference chemicals to freshly isolated rainbow trout hepatocytes and *Daphnia magna*. Aquat Toxicol 30:47–60

Lin AY-C, Yu T-H, Lateef SK (2009) Removal of pharmaceuticals in secondary wastewater treatment processes in Taiwan. J Hazard Mater. doi:10.1016/j.Jhazmat.2009.01.108

Lister A, Regan C, Van Zwol J, Van Der Kraak G (2009) Inhibition of egg production in zebrafish by fluoxetine and municipal effluents: a mechanistic evaluation. Aquat Toxicol 95:320–329

Long ER, Chapman PM (1985) A sediment quality triad-measures of sediment contamination, toxicity and infaunal community composition in Puget Sound. Mar Environ Poll 16:405–415

Lundholm CE (1997) DDE-induced eggshell thinning in birds: effects of p, p'-DDE on the calcium and prostaglandin metabolism of the eggshell gland. Comp Biol Physiol 118C:113–128

Macri A, Sbardella E (1984) Toxicological evaluation of nitrofurazone and furazolidone on *Selenaslrum capricornurum*, *Daphnia magna* and *Musca domes rica*. Ecotoxicol Environ Saf 8:115–105

Majewski AR, Blanchfield PJ, Wautier PK (2002) Waterborne 17alfa-ethynylestradiol affects aggressive behavior of male fathead minnows (*Pimephales promelas*) under artificial spawning conditions. Water Qual Res J Can 37:697–710

Manocha S, Venkataraman S (2000) Pharmacological and histopathological evaluation of ulcer formation and end organ toxicity by NSAIDS with concurrent ranitidine treatment in aged rats. In: Presentation at sixth Internet World Congress for Biomedical Science

Manuell A, Beligni MV, Yamaguchi K, Mayfield SP (2004) Regulation of chloroplast translation: interactions of RNA elements, RNA-binding proteins and the plastid ribosome. Biochem Soc Trans 23:601–605

Marking L, Howe GE, Crowther JR (1988) Toxicity of erythromycin, oxytetracycline and tetracycline administered to Lake Trout in water baths, by injection or by feeding. The Progressive Fish-Culturist 50:197–201

Marques CR, Abrantes N, Goncaes F (2004) Life-history traits of standard and autochthonous cladocerans: II. Acute and chronic effects of acetylsalicylic acid metabolites. Environ Toxicol 19:527–540

Martinovic D, Hogarth WT, Jones RE, Sorensen PW (2007) Environmental estrogens suppress hormones, behavior, and reproductive fitness in male fathead minnows. Environ Toxicol Chem 26:271–278

Masuda S, Deguchi Y, Masuda Y, Watanabe T, Nukaya H, Terao Y, Takamura T, Wakabayashi K, Kinae N (2004) Genotoxicity of 2-[2-(acetylamino)-4-[bis(2-hydroxyethyl)amino]-5-methoxyphenyl]-5-amino-7-bromo-4-chloro-2 H-benzotriazole (PBTA-6) and 4-amino-3,3′-dichloro-5,4′-dinitro-biphenyl (ADDB) in goldfish (*Carassius auratus*) using the micronucleus test and the comet assay. Mutat Res 560:33–40

McFadden GI, Roos DS (1999) Apicomplexan plastids as drug targets. Trends Microbiol 7:328–333

McMurry LM, Oethinger M, Levy SB (1998) Triclosan targets lipid synthesis. Nature 394: 531–532

Meguid MM, Fetissov SO, Varma M, Sato T, Zhang L, Laviano A, Rossi-Fanelli F (2000) Hypothalamic dopamine and serotonin in the regulation of food intake. Nutrition 16:843–857

Mehinto AC, Hill EM, Tyler CR (2010) Uptake and biological effects of environmentally relevant concentrations of the nonsteroidal anti-inflammatory pharmaceutical diclofenac in rainbow trout (*Oncorhynchus mykiss*). Environ Sci Technol 44:2176–2182

Meinertz JR, Schreier M, Bernardy JA, Franz JL (2010) Chronic toxicity of diphenhydramine hydrochloride and erythromycin thiocyanate to daphnia, *Daphnia magna*, in a continuous exposure test system. Bull Environ Contam Toxicol 85:447–451

Mennigen JA, Sassine J, Trudeau VL, Moon TW (2010a) Waterborne fluoxetine disrupts feeding and energy metabolism in the goldfish *Carassius auratus*. Aquat Toxicol 100:128–137

Mennigen JA, Lado WE, Zamora JM, Duarte-Guterman P, Langlois VS, Metcalfe CD, Chang JP, Moon TW, Trudeau VL (2010b) Waterborne fluozetine disrupts the reproductive axis in sexually mature male goldfish, *Carassius auratus*. Aquat Toxicol 100:354–364

Mennigen JA, Stroud P, Zamora JM, Moon TW, Trudeau VL (2011) Pharmaceuticals as neuroendocrine disruptors: lessons learned from fish on Prozac. J Toxicol Environ Health, Part B 14:387–412

Miller DH, Jensen KM, Villeneuve DL, Kahl MD, Makynen EA, Durhan EJ, Ankley GT (2007) Linkage of biochemical responses to population-level effects: a case study with vitellogenin in the fathead minnow (*Pimephales promelas*). Environ Toxicol Chem 26:521–527

Mimeault C, Woodhouse AJ, Miao XS, Metcalfe CD, Moon TW, Trudeau VL (2005) The human lipid regulator, gemfibrozi bioconcentrates and reduces testosterone in the goldfish, *Carassius auratus*. Aquat Toxicol 73:44–54

Minagh E, Hernan R, O'Rourke K, Lyng FM, Davoren M (2009) Aquatic ecotoxicity of the selective serotonin reuptake inhibitor sertraline hydrochloride in a battery of freshwater test species. Ecotoxicol Environ Saf 72:434–440

Mitchelmore CL, Chipman JK (1998a) Detection of DNA strand breaks in Brown trout (*Salmo trutta*) hepatocytes and blood cells using the single cell gel electrophoresis (comet) assay. Aquat Toxicol 41:161–182

Mitchelmore CL, Chipman JK (1998b) DNA strand breakage in aquatic organisms and the potential value of the comet assay in environmental monitoring. Mutat Res 399:135–147

Mitchelmore CL, Birmelin C, Livingstone DR, Chipman JK (1997) Polycyclic and nitro-aromatic compounds produce DNA strand breakage in Brown trout (*Salmo trutta*) and mussel (*Mytilus edulis* L.) cells. Mutagenesis 12:101

Monteiro SC, Boxall ABA (2010) Occurrence and fate of human pharmaceuticals in the environment. Rev Environ Contam Toxicol 202:53–154

Moore MT, Greenway SL, Farris JL, Guerra B (2008) Assessing caffeine as an emerging environmental concern using conventional approaches. Arch Environ Contam Toxicol 54:31–35

Mu X, LeBlanc GA (2002) Developmental toxicity of testosterone in the crustacean *Daphnia magna* involves anti-ecdysteroidal activity. Gen Comp Endocrinol 129:127–133

Nakamura Y, Yamamoto H, Sekizawa J, Kondo T, Hirai N, Tatarazako N (2008) The effects of pH on fluoxetine in Japanese medaka (*Oryzias latipes*): acute toxicity in fish larvae and bioaccumulation in juvenile fish. Chemosphere 70:865–873

Nassef M, Matsumoto S, Seki M, Khalil F, Kang IJ, Shimasaki Y, Oshima Y, Honjo T (2010) Acute effects of ticlosan, diclofenac and carbamazepine on feeding performance of Japanese medaka fish (*Oryzias latipes*). Chemosphere 80:1095–1100

Nation JL (2002) Insect physiology and biochemistry. CRC Press, Boca Raton

Nayler WG, Poole-Wilson CH (1981) Calcium antagonists: definition and mode of action. Basic Res Cardiol 76:1–15

Nentwing G (2007) Effects of pharmaceuticals on aquatic invertebrates. Part II: the antidepressant drug fluoxetine. Arch Environ Contam Toxicol 52:163–170

Nentwing G, Oetken M, Oehlmann J (2004) Effects of pharmaceuticals on aquatic invertebrates – the example of carbamazepine and clofibric acid. In: Kummerer K (ed) Pharmaceuticals in the environment: sources, fate, effects and risks, 2nd edn. Springer, Berlin, Germany, pp 195–208

Neuman NF, Galvez F (2002) DNA microarrays and toxicogenomics: applications for ecotoxicology? Biotechnol Adv 20:391–419

Nickerson JG, Gugan SG, Drouin G, Moon TW (2001) A putative β_2-adrenoceptor from the rainbow trout (*Oncorhyncus mykiss*). Molecular characterixation and pharmacology. Eur J Biochem 268:6465–6472

Nie X, Gu J, Lu J, Pan W, Yang Y (2009) Effects of norfloxacin and butylated hydroxyanisole on the freshwater microalga *Scenedesmus obliquus*. Ecotoxicology 18:677–684

Oakes KD, Coors A, Escher BI, Fenner K, Garric J, Gust M, Knacker T, Küster A, Kussatz C, Metcalfe CD, Monteiro S, Moon TW, Mennigen JA, Parrott J, Péry ARR, Ramil M, Roennefahrt I, Tarazona JV, Sánchez-Argüello P, Ternes TA, Trudeau VL, Boucard T, Van Der Kraak GJ, Servos MR (2010) Environmental risk assessment for the serotonin re-uptake inhibitor fluoxetine: case study using the European risk assessment framework. Integr Environ Assess Manag 6(Suppl 1):524–539

Oaks JL, Gilbert M, Virani MZ, Watson RT, Meteyer CU, Rideout BA, Shivaprasad HL, Ahmed S, Chaudhry MJI, Arshad M, Mahmood S, Ali A, Khan AA (2004) Diclofenac residues as the cause of vulture population decline in Pakistan. Nature 427:630–633

OECD (1999) Detailed review paper: appraisal of test methods for sex-hormone disrupting chemicals. OECD Environmental Health and Safety Publication No. 21

Orn S, Holbech H, Madsen TH, Norrgren L, Petersen GI (2003) Gonad development and vitellogenin production in zebrafish (*Danio rerio*) exposed to ethinylestradiol and methyltestosterone. Aquat Toxicol 65:397–411

Owen SF, Giltrow E, Huggett DB, Hutchinson TH, Saye J, Winter MJ, Sumpter JP (2007) Comparitive physiology, pharmacology and toxicology of β-blockers: mammals versus fish. Aquat Toxicol 82:145–162

Painter MM, Buerkley MA, Julius ML, Vajda AM, Norris DO, Barber LB, Furlong ET, Schultz MM, Schoenfuss HL (2009) Antidepressants at environmentally relevant concentrations affect predator avoidance behavior of larval fathead minnows (*Pimephales promelas*). Environ Toxicol Chem 28:2677–2684

Panter GH, Thompson RS, Beresford N, Sumpter JP (1999) Transformation of a non-oestrogenic steroid metabolite to an oestrogenically active substance by minimal bacterial activity. Chemosphere 38(15):3579–3596

Panter GH, Hutchinson TH, Hurd KS, Sherren A, Stanley RD, Tyler CR (2004) Successful detection of (anti-)androgenic and aromatase inhibitors in pre-spawning adult fathead minnows (*Pimephales promelas*) using easily measured end points of sexual development. Aquat Toxicol 70:11–21

Parrott JL, Blunt BR (2005) Life-cycle exposure of fathead minnows (*Pimephales promelas*) to an ethinylestradiol concentration below 1 ng/L reduces egg fertilization success and demasculinize males. Environ Toxicol 20:131–141

Parrott JL, Bennie DT (2009) Life-cycle exposure of fathead minnows to a mixture of six common pharmaceuticals and triclosan. J Toxicol Environ Health A 72:633–641

Pascoe D, Karntanut W, Muller CT (2003) Do pharmaceuticals affect freshwater invertebrates? A study with the cnidarians *Hydra vulgaris*. Chemosphere 51:521–528

Pawlowski S, van Aerle R, Tyler CR, Braunbeck T (2004) Effects of 17α-ethinylestradiol in a fathead minnow (*Pimephales promelas*) gonadal recrudescence assay. Ecotoxicol Environ Saf 57:330–345

Perreault HA, Semsar K, Godwin J (2003) Fluoxetine treatment decreases territorial aggression in a coral reef fish. Physiol Behav 79(4–5):719–724

Pery ARR, Gust M, Vollat B, Mons R, Ramil M, Fink G, Ternes T, Garric J (2008) Fluoxetine effects assessment on the life cycle of aquatic invertebrates. Chemosphere 73:300–304

Pomati F, Netting AG, Calomari D, Neilan BA (2004) Effects of erythromycin, tetracycline and ibuprofen on the growth of *Synechocystis* sp. and *Leman minor*. Aquat Toxicol 67:387–396

Pounds N, Maclean S, Webley M, Pascoe D, Hutchinson T (2008) Acute and chronic effects of ibuprofen in the mollusk *Planorbis carnatus* (*Gastropoda: Planorbidae*). Ecotoxicol Environ Saf 70:47–52

Powers DA (1989) Fish as model systems. Science 246:352–358

Pro J, Ortiz JA, Bolocs S, Fernandez C, Carbonell G, Tarazona JV (2003) Effect assessment of antimicrobial pharmaceuticals on the aquatic plant *Lemna minor*. Bull Environ Contam Toxicol 70:290–295

Purdom CE, Hardiman PA, Bye VJ, Eno NC, Tyler CR, Sumpter JP (1994) Estrogenic effects of effluents from sewage treatment works. Chem Ecol 8:275–285

Quinn B, Gagne F, Blaise C (2008) An investigation into the acute and chronic toxicity of eleven pharmaceuticals (and their solvents) found in wastewater effluent on the cnidarians, *Hydra attenuate*. Sci Total Environ 389:306–314

Ralph S, Petras M, Pandrangi R, Vrzoc M (1996) Alkaline single cell gel (comet) assay and geno-toxicity monitoring using two species of tadpoles. Environ Mol Mutagen 28:112–120

Ramirez AJ, Mottaleb MA, Brooks BW, Chambliss CK (2007) Analysis of pharmaceuticals in fish using liquid chromatography-tandem mass spectrometry. Anal Chem 79:3155–3163

Ramirez AJ, Brain RA, Usenko S, Mottaleb MA, O'Donnell JG, Stahl LL, Wethen JB, Snyder BD, Pitt JL, Perez-Hurtado P, Dobbins LL, Brooks BW, Chambliss CK (2009) Occurrence of phar-maceuticals and personal care products in fish: results of a national pilot study in the United States. Environ Toxicol Chem 28:2587–2597

Rand GM (ed) (1995) Fundamentals of aquatic toxicology, 2nd edn. Taylor and Francis, Philadelphia, PA

Revai T, Harmos G (1999) Nephrotic syndrome and acute interstitial nephritis associated with the use of diclofenac. Wien Klin Wochenschr 111:523–524

Richards SM, Cole SE (2006) A toxicity and hazard assessment of fourteen pharmaceutical com-pounds to *Xnopus laevis* larvae. Ecotoxicology 15:647–656

Robinson AA, Belden JB, Lydy MJ (2005) Toxicity of fluoroquinolone antibiotics to aquatic organisms. Environ Toxicol Chem 24:423–430

Rodriguez C, Chellman K, Gomez S, Marple L (1992) Environmental assessment report pursuant to 21 CFR 25.31(a) submitted to the US FDA in support of the New Drug Application (NDA) for naproxen for over-the-counter use. Hamilton Pharmaceuticals Limited, Puerto Rico

Rosal R, Rodea-Palomares I, Boltes K, Fernandez-Pinas F, Leganes F, Gonzalo S, Pere A (2010) Ecotoxicity assessment of lipid regulators in water and biologically treated wastewater using three aquatic organisms. Environ Sci Pollut Res Int 17:135–144

Runnalls TJ, Hala DN, Sumpter JP (2007) Preliminary studies into the effects of the human pharmaceu-tical clofibric acid on sperm parameters in adult fathead minnow. Aquat Toxicol 84:111–118

Russom CL, Bradbury SP, Broderius SJ, Hammermeister DE, Drummond RA (1997) Predicting modes of toxic action from chemical structure: acute toxicity in the fathead minnow (*Pimephales promelas*). Environ Toxicol Chem 16:948–967

Saito H, Sudo M, Shigeoka T, Yamauchi F (1991) In vitro cytotoxicity of chlorophenols to goldfish GF-Scale (GFS) cells an quantitative structure-activity relationships. Environ Toxicol Chem 10:235–241

Sanderson H, Thomsen M (2007) Ecotoxicological quantitative structure activity relationships for pharmaceuticals. Bull Environ Contam Toxicol 79:331–335

Sanderson H, Thomsen M (2009) Comparative analysis of pharmaceuticals versus industrial chemicals acute aquatic toxicity classification according to the United Nations classification system for chemicals: assessment of the (Q)SAR predictability of pharmaceuticals acute aquatic toxicity and their predominant acute toxic mode-of-action. Toxicol Lett 187:84–93

Sanderson H, Johnson DJ, Wilson CJ, Brain RA, Solomon KR (2003) Probabilistic hazard assessment of environmentally occurring pharmaceuticals toxicity to fish, daphnids and algae by ECOSAR screening. Toxicol Lett 144:383–395

Sanderson H, Johnson DJ, Reitsma T, Brain RA, Wilson CJ, Solomon KR (2004) Ranking and prioritization of environmental risks of pharmaceuticals in surface waters. Regul Toxicol Pharmacol 39:158–183

Sanofi (1996) Tiludronate disodium material safety data sheet. SR 41319B. Sanofi Research

Santos L, Araujo AN, Fachini A, Pena A, Delerue-Matos C, Montenegro MC (2010) Ecotoxicological aspects related to the presence of pharmaceuticals in the aquatic environment. J Hazard Mater 175:45–95

Sanyal AK, Roy D, Chowdhury B, Banerjee AB (1993) Ibuprofen, a unique anti-inflammatory compound with antifungal activity against dermatophytes. Lett Appl Microbiol 17:109–111

Schena M, Shalon D, Davis RW, Brown PO (1995) Quantitative monitoring of gene expression patterns with a complimentary DNA microarray. Science 270:467–470

Schmidt W, O'Rourke K, Hernan R, Quinn B (2011) Effects of pharmaceuticals gemfibrozil and diclofenac on the marine mussle (*Mytilus* spp.) and their comparison with standardized toxicity tests. Mar Pollut Bull 62:1389–1395

Schmitt H, Bouchard T, Garric J, Jensen J, Parrott J, Pery A, Rombke J, Straub JO, Hutchinson TH, Sanchez-Arguello P, Wennmalm A, Duis K (2010) Recommendations on the environmental risk assessment of pharmaceuticals: effect characterization. Integr Environ Assess Manag 6(Suppl 1):588–602

Schnell S, Bols NC, Barata C, Porte C (2009) Single and combined toxicity of pharmaceuticals and personal care products (PPCPs) on the rainbow trout liver cell line RTL-W1. Aquat Toxicol 93:244–252

Schulte-Oehlmann U, Oetken M, Bachmann J, Oehlmann J (2004) Effects of ethinyloestradiol and methyltestosterone in prosobranch snails. In: Kummerer K (ed) Pharmaceuticals in the environment: sources, fate, effects and risks, 2nd edn. Springer, Berlin, Germany, pp 233–247

Schultz MM, Furlong ET, Kolpin DW, Werner SL, Schoenfuss HL, Barber LB, Blazer VS, Norris DO, Vajda AM (2010) Antidepressant pharmaceuticals in two U.S. effluent-impacted streams: occurrence and fate in water and sediment, and selective uptake in fish neural tissue. Environ Sci Technol 44:1918–1925

Schwaiger J, Ferling H, Mallow U, Wintermayr H, Negele RD (2004) Toxic effects of the non-steroidal anti-inflammatory drug diclofenac. Part I: histopathological alterations and bioaccumulation in rainbow trout. Aquat Toxicol 68:141–150

Schweinfurth H, Lange R, Schneider PW (1996) Environmental risk assessment in the pharmaceutical industry. In: Presentation at the 3 rd Eurolab Symposium: testing and analysis for industrial competitiveness and sustainability. Berlin, Germany

Segner H, Caroll K, Fenske M, Janssen CR, Maack G, Pascoe D, Schäfers C, Vendenbergh GF, Watts M, Wenzel A (2003a) Identification of endocrine disrupting effects in aquatic vertebrates and invertebrates: report from the European IDEA project. Ecotoxicol Environ Saf 54:302–314

Segner H, Navas JM, Schäfers C, Wenzel A (2003b) Potencies of estrogenic compounds in in vitro screening assays and in life cycle tests with zebrafish in vivo. Ecotoxicol Environ Saf 54:315–322

Seki M, Yokota H, Matsubara H, Tsuruda Y, Maeda M, Tadokoro H, Kobayashi K (2002) Effect of ethinylestradiol on the reproduction and induction of vitellogenin and testis-ova in medaka (*Oryzias latipes*). Environ Toxicol Chem 21:1692–1698

Semsar K, Perreault HA, Godwin J (2004) Fluoxetine-treated male wrasses exhibit low AVT expression. Brain Res 1029(2):141–147

Sheahan DA, Bucke D, Matthiessen P, Sumpter JP, Kirby MF, Neall P, Waldock M (1994) The effects of low levels of 17α-ethinylestradiol upon plasma vitellogenin levels in male and female rainbow trout, *Oncorhynchus mykiss*, held at two acclimation temperatures. In: Müller R, Lloyd R (eds) Sublethal and chronic effects of pollutants on freshwater fish. Blackwell, Oxford, UK, pp 99–112

Singh NP, McCoy MT, Tice RR, Schneider EL (1998) A simple technique for quantification of low levels of DNA damage in individual cells. Exp Cell Res 175:184–191

Smith GR, Burgett AA (2005) Effects of three organic wastewater contaminants on American toad, *Bufo americanus*, tadpoles. Ecotoxicology 14:477–482

Sole M, Shaw JP, Frickers PE, Readman JW, Hutchinson TH (2010) Effects on feeding rate and biomarker responses of marine mussels experimentally exposure to propranolol and acetaminophen. Anal Bioanal Chem 396:649–656

Staels B, Dallongeville J, Auwerx J, Schoonjans K, Leitersdorf E, Fruchart J-C (1998) Mechanism of action of fibrates on lipid and lipoprotein metabolism. Circulation 98:2088–2093

Stanley JK, Brooks BW (2009) Perspectives on ecological risk assessment of chiral compounds. Integr Environ Assess Manag 5:364–373

Stanley JK, Ramirez AJ, Mottaleb M, Chambliss CK, Brooks BW (2006) Enantio-specific toxicity of the β-blocker propranolol to *Daphnia magna* and *Pimephales promelas*. Environ Toxicol Chem 25:1780–1786

Stanley JK, Ramirez AJ, Chambliss CK, Brooks BW (2007) Enantiospecific sublethal effects of the antidepressant fluoxetine to a model aquatic vertebrate and invertebrate. Chemosphere 69:9–16

Stuer-Lauridsen F, Birkved M, Hansen LP, Lutzhoft HC, Halling-Sorensen B (2000) Environmental risk assessment of human pharmaceuticals in Denmark after normal therapeutic use. Chemosphere 40:783–793

Sumpter JP, Jobling S (1995) Vitellogenesis as a biomarker for estrogenic contamination of the aquatic environment. Environ Health Perspect 103(Suppl 7):173–178

Sweetman SC (ed) (2002) Martindale: the complete drug reference, 33rd edn. Pharmaceutical, London

Ternes T, Jos A, Siegrist H (2004) Scrutinizing pharmaceutical and personal care products in wastewater treatment. Environ Sci Technol 38:393–399

Thompson RJ, Mosig G (1985) An ATP-dependant supercoiling topoisomerase of *Chlamydomonas reinhardtii* affects accumulation of specific chloroplast transcripts. Nucleic Acids Res 13:873–891

Tooby TE, Hursey PA, Alabaster JS (1975) The acute toxicity of 102 pesticides and miscellaneous substances to fish. Chem Ind 611975:523–526

Triebskorn R, Casper H, Heyda A, Eikemper R, Kohler H-R, Schwaiger J (2004) Toxic effects of the non-steroidal anti-inflammatory drug diclofenac. Part II. Cytological effects in liver, kidney, gills, and intestine of rainbow trout (*Oncorhynchus mykiss*). Aquat Toxicol 68:176–183

Triebskorn R, Casper H, Scheil V, Schwaiger J (2007) Ultrastructural effects of pharmaceuticals (carbamezapine, clofibric acid, metoprolol, diclofenac) in rainbow trout (*Oncorhynchus mykiss*) and common carp (*Cyprinus carpio*). Anal Bioanal Chem 387:1405–1416

Ullman U, Svedmyr N (1988) Salmeterol, a new long acting inhaled beta2 adrenoreceptor agonist: comparison with salbutamol in adult asthmatic patients. Thorax 43:674–678

United States Environmental Protection Agency (1994) Policy for the development of effluent limitations in National Pollutant Discharge Elimination System (NPDES) permits to control whole effluent toxicity for the protection of aquatic life. EPA833-B-94-002. Office of Water, Washington, DC

United States Environmental Protection Agency (2000) Method guidance and recommendations for whole effluent toxicity (WET) testing (40 CFR Part 136). EPA/821/B-00-004. Office of Water, Office of Science and Technology. Washington DC

United States Environmental Protection Agency (2002a) Methods for measuring the acute toxicity of effluents and receiving waters to freshwater and marine organisms. EPA 821/R-02/012, 5th edn. Office of Science and Technology, Washington, DC

United States Environmental Protection Agency (2002b) Short-term methods for estimating the chronic toxicity of effluents and receiving waters to freshwater organisms. EPA 821/R-02-013, 4th edn. Office of Science and Technology, Washington, DC

United States Environmental Protection Agency (2002c) Short-term methods for estimating the chronic toxicity of effluents and receiving waters to marine and estuarine organisms, 3rd edn. Office of Science and Technology, Washington DC

Valenti TW, Perez Hurtado P, Chambliss CK, Brooks BW (2009) Aquatic toxicity of sertraline to *Pimephales promelas* at environmentally relevant surface water pH. Environ Toxicol Chem 28:2685–2694

Valenti TW, Taylor JT, Back JA, King RS, Brooks BW (2011) Influence of drought and total phosphorus on diel pH in wadeable streams: implications for ecological risk assessment of ionizable contaminants. Integr Environ Assess Manag 7:636–647

Valenti TV, Gould GG, Berninger JP, Connors KA, Keele NB, Prosser KN, Brooks BW (2012) Human therapeutic plasma levels of the selective serotonin reuptake inhibitor (SSRI) sertraline decrease serotonin reuptake transporter binding and shelter seeking behavior in adult male fathead minnows. Environ Sci Technol 46:2427–2435

van den Brandof E, Montforts M (2010) Fish embryo toxicity of carbamazepine, diclofenac and meoprolol. Ecotoxicol Environ Saf 73:1862–1866

van der Ven K, De Wit M, Keil D, Moens L, Van Leeriput K, Naudts B, De Coen W (2005) Development and application of a brain-specific cDNA microarray for effect evaluation of neuroactive pharmaceuticals in zebrafish (*Danio rerio*). Comp Biochem Physiol B 141:408–417

van Zwieten PA (1994) Amlodipine: an overview of its pharmacodynamic and pharmacokinetic properties. Clin Cardiol 17(Suppl 3):1113–1116

Vandenbergh GF, Adriaens D, Verslycke T, Janssen CR (2003) Effects of 17α-ethinylestradiol on sexual development of the amphipod *Hyalella azteca*. Ecotoxicol Environ Saf 54:216–222

Vane JR, Botting RM (1998) Mechanism of action of anti-inflammatory drugs. Int J Tissue React 20:3–15

VICH (2000) Environmental impact assessment (EIAs) for veterinary medicinal products (Veterinary medicines) – Phase 1. VICH GL6 (Ecotoxicity Phase 1), June 2000, For Implementation at Phase 7

VICH (2004) Environmental impact assessment for veterinary medicinal products phase II guidance, VICH-GL38. Brussels. Belgium, VICH, p 38, http://vich.eudra.org/

Vos JG, Dybing E, Greim H, Ladefoged O, Lambre C, Tarazona JV, Brandt I, Vethaak AD (2000) Health effects of endocrine-disrupting chemicals on wildlife with special reference to the European situation. Crit Rev Toxicol 30:71–133

Wall MK, Mitchall LA, Maxwell A (2004) *Arabidopsis thaliana* DNA gyrase is targeted to chloroplasts and mitochondria. Proc Natl Acad Sci USA 101:7821–7826

Walton RJ, Sherif IT, Noy GA, Alberti KG (1979) Improved metabolic profiles in insulin-treated diabetic patients given an alpha-glucoside hydrolase inhibitor. Br Med J 6158:220–221

Wang WH, Lay JP (1989) Fate and effects of salicylic acid compounds in freshwater systems. Ecotoxicol Environ Saf 17(3):308–316

Watts MM, Pascoe D, Carroll K (2002) Population responses of the freshwater amphipod *Gammarus pulex* (L.) to an environmental oestrogen 17α ethinylestradiol. Environ Toxicol Chem 21:445–450

Watts MM, Pascoe D, Carroll K (2003) Exposure to 17α ethinylestradiol and bisphenol A-effects on larval moulting and mouthpart structure of *Chironomus riparius*. Ecotoxicol Environ Saf 54:207–215

Webb SF (2001) A data based perspective on the environmental risk assessment of human pharmaceuticals II: aquatic risk characterisation. In: Kümmerer K (ed) Pharmaceuticals in the environment. Sources, fate, effects and risks. Springer, Berlin, pp 319–343

Welborn TL (1969) The toxicity of nine therapeutic and herbicidal compounds to striped bass. The Progressive Fish Culturist 31(1):27–32

Weston A, Caminada D, Galicia H, Fent K (2009) Effects of lipid-lowering pharmaceuticals bezafibrate and clofibric acid on lipid metabolism in fathead minnow (*Pimephales promelas*). Environ Toxicol Chem 28:2648–2655

Wilford WA (1966) Toxicity of 22 therapeutic compounds to six fishes. US Dept. of the Interior, Fish and Wildlife Service, Bureau of Sports Fisheries and Wildlife, Washington DC (Resource Publication 35)

Williams EMV (1958) The mode of action of quinidine on isolated rabbit atria interpreted from intracellular potential records. Br J Pharmacol Chemother 13:276–287

Winter MJ, Lillicrap AD, Caunter JE, Schaffner C, Alder AC, Ramil M, Ternes TA, Giltrow E, Sumpter JP, Hutchinson TH (2008) Defining the chronic impacts of atenolol on embryo-larval development and reproduction in the fathead minnow (Pimephales promelas). Aquat Toxicol 86:361–369

Wong DT, Perry KW, Bymaster FP (2005) The discovery of fluoxetine hydrochloride (Prozac). Nat Rev Drug Discov 4:764–774

Yamashita N, Yasojima M, Miyajima K, Suzuki Y, Tanaka H (2006) Effects of antibacterial agents, levofloxacin and clarithromycin, on aquatic organisms. Water Sci Technol 53:65–72

Zeilinger J, Steger-Hartmann T, Maser E, Goller S, Vonk R, Lange R (2009) Effects of synthetic gestagens on fish reproduction. Environ Toxicol Chem 28:2663–2670

Zerulla M, Lange R, Steger-Hartmann T, Panter G, Hutchinson T, Dietrich DR (2002) Morphological sex reversal upon short-term exposure to endocrine modulators in juvenile fathead minnow (Pimephales promelas). Toxicol Lett 131:51–63

Zhang X, Oakes KD, Cui SF, Bragg L, Servos MR, Pawliszyn J (2010) Tissue-specific in vivo bioconcentration of pharmaceuticals in rainbow trout (Oncorhynchus mykiss) using space-resolved solid-phase microextraction. Environ Sci Technol 44:3417–3422

Zhou SN, Oakes KD, Servos MR, Pawliszyn J (2008) Application of solid-phase microextraction for in vivo laboratory and field sampling of pharmaceuticals in fish. Environ Sci Technol 42:6073–6079

Zou E, Fingerman M (1997) Synthetic estrogenic agents do not interfere with sex differentiation but do inhibit molting of the cladoceran Daphnia magna. Bull Environ Contam Toxicol 58:596–602

Zuccato E, Castiglioni S, Fanelli R, Bagnati R, Reitano G, Calamari D (2004) Risks related to the discharge of pharmaceuticals in the environment: further research is needed. In: Kümmerer K (ed) Pharmaceuticals in the environment: sources, fate, effects and risks, 2nd edn. Springer, Berlin, Germany, pp 431–437

Zurita JL, Repetto G, Jos A, Salguero M, Lopez-Artiguez M, Camean AM (2007) Toxicological effects of the lipid regulator gemfibrozil in four aquatic systems. Aquat Toxicol 81:106–115

Fish Metalloproteins as Biomarkers of Environmental Contamination

Rachel Ann Hauser-Davis, Reinaldo Calixto de Campos, and Roberta Lourenço Ziolli

Contents

R.A. Hauser-Davis (✉) • R.C. de Campos
Departamento de Química, Pontifícia Universidade Católica do Rio de Janeiro (PUC-Rio),
Rua Marquês de São Vicente, 225, Gávea, CEP: 22453-900, Rio de Janeiro, RJ, Brazil

Instituto Nacional de Ciência e Tecnologia - INCT de Bioanalítica, CNPq,
Rio de Janeiro, RJ, Brazil
e-mail: rachel.hauser.davis@gmail.com; rccampos@puc-rio.br

R.L. Ziolli
Instituto de Bioiências, Universidade Federal do Estado do Rio de Janeiro (UNIRIO),
Av. Pasteur, 458, Urca, CEP: 22290-240, Rio de Janeiro, RJ, Brazil
e-mail: rlziolli@puc-rio.br

D.M. Whitacre (ed.), *Reviews of Environmental Contamination and Toxicology*,
Reviews of Environmental Contamination and Toxicology 218,
DOI 10.1007/978-1-4614-3137-4_2, © Springer Science+Business Media, LLC 2012

1   Introduction

In this review, we explore the fish metalloproteins that have been discovered by 'omic techniques, and their application as fish biomarkers of environmental contamination.

The fields collectively known as "'omics" have undergone tremendous development in the past decade. The best known among them is genomics, in which complete genome DNA sequences of living organisms are produced. Other "'omics" include structural genomics, proteomics, toxicoproteomics, and metallomics (Shi and Chance 2008). The information obtained by applying the fields of proteomics, toxicoproteomics, and metallomics can be utilized to establish biomarkers of exposure for organisms that are affected by environmental contaminants. Moreover, the likelihood of discerning viable biomarkers from proteomic and metallomic investigations is especially high, given that protein profiles appear to be specific to particular stressors (Martin et al. 2001; Shepard et al. 2000; Jellum et al. 1983; Blom et al. 1992; Powers 1989). Thus, environmental proteomics provides a more comprehensive assessment of the toxic and defensive mechanisms that are triggered by pollutants than do traditional biomarker studies (Gonzalez-Fernandez et al. 2008).

For several reasons, fish have attracted considerable interest in studies designed to assess the biological and biochemical responses that organisms have to environmental contaminants (Powers 1989). Fish are particularly useful for assessing water-borne and sediment-deposited toxins, and may provide advanced warning of the environmental contamination potential of new chemicals, or the status of environmental contamination by well-known toxicants. Fish are also particularly good models for studies in which biochemistry and comparative physiology are involved, because they live in diverse habitats and must adapt to environmental parameters and stress, both of which can be easily reproduced under laboratory conditions (Beyer 1996). The understanding of toxicant uptake, behavior, and responses in fish, therefore, has a high potential for ecological relevance. Because of these aspects, it is important that these organisms be studied in more detail; the development of 'omic techniques in recent years offers the possibility to perform such studies in new and useful ways.

2   Biomarkers and Proteomic Approaches to Their Discovery

Biomarkers have been extensively studied, and several comprehensive reviews and books on the subject are available, including ones that address biomarkers specific to fish (Adams 1987; NRC 1987; Schlenk 1999; Stegeman et al. 1992). Biomarkers are defined as measurements in body fluids, cells, or tissues that can indicate biochemical or cellular modifications resulting from the presence of toxicants or stress (NRC 1987). This original definition was later modified to take into account characteristics of organisms, populations, or communities, including behavior, in which measurable responses occur that reflect changes to the environment (Adams 1987; Depledge et al. 1992). In other words, the foundational concept of the biomarker approach for assessing adverse effects or stress is based on the hypothesis that the effects of stress are typically manifested, first at lower levels of biological organization, before disturbances are realized at the population-, community-, or ecosystem-levels. This concept allows for development of early warning biomarker effect signals that may occur at later response levels (Bayne et al. 1985).

The emerging omics technologies, are well positioned to address such concerns by fomenting the discovery of mechanisms that underlay the toxic action of chemical pollutants, and by assisting in the identification of new biomarkers (Dowling and Sheehan 2006; López-Barea and Gómez-Ariza 2006; Quackenbush 2001). For example, biomarkers may consist of an integrated set of genes or proteins that are simultaneously expressed in certain situations. Biomarkers may be used to (1) characterize related functions of genes and gene products that have similar activity profiles or common mechanisms of regulation (Snape et al. 2004); (2) classify compounds that have similar modes of action by creating toxicological "fingerprints" for them (Oberemm et al. 2005; Miracle and Ankley 2005; Baker 2005); or (3) characterize different stress levels by integrating general and highly specific markers in one assay. Thus, the incredible amount of molecular-level information available from applying 'omic-technologies is rapidly fomenting the development of biomarkers, from singular biomarker measurements to highly complex multi-marker genomic/proteomic panels (Miracle and Ankley 2005).

Proteomic analyses provide valuable information, when variations that occur within the proteome of organisms are compared as a consequence of biological perturbations or external stimuli. These stimuli often result in different protein expressions or the redistribution of specific proteins within cells (Martin et al. 2001, 2003; Vilhelmsson et al. 2004; Tyers and Mann 2003), which can be correlated with environmental contamination, and which may help identify proteins that are altered from pollutant exposure, or may help establish a protein-pollutant toxic mechanism relationship (López-Barea and Gómez-Ariza 2006). An added bonus in these studies is the fact that it is not absolutely necessary to establish the identity of a protein for it to become a successful biomarker of exposure. Indeed, the characteristics of a peptide and the specific conditions under which it occurs are the more pressing concerns (Hogstrand et al. 2002). Recent studies have produced protein expression signatures that were characterized in marine invertebrates, in response to changing salinities and temperatures, and in response to the presence of polychlorinated biphenyls and copper (Bradley et al. 1985; Shepard et al. 2000; Shepard and Bradley 2000; Kimmel and Bradley 2001).

3   Fish Proteomics in the Search for Biomarkers of Environmental Contamination

Functional genomic and proteomic technologies have allowed biological questions to be addressed in a broader approach by allowing simultaneous study of thousands of genes or proteins.

Applications of comparative proteome "'omic" techniques have proven useful for further examining the altered protein expression in fish development. These techniques have enhanced, for example, the understanding of the molecular mechanisms underlying normal and abnormal development (Kultz and Somero 1996; Kanaya et al. 2000), the effects of feeding habits on metabolism parameters and the liver proteome (Martin et al. 2003; Vilhelmsson et al. 2004), and the understanding of physiological mechanisms that explain the nature of phenotypic differences between farmed and wild fish (Olsson et al. 2007).

Comparing fish proteomes in different situations is appealing, because changes in proteome expression under complex field situations may disclose which gene products, metabolites, or proteins are most interesting to investigate (Albertsson et al. 2007). Such comparative proteomic studies are also useful as tools in the investigation of fish development and different ecological situations.

Within this context, Tay et al. (2006) used two-dimensional electrophoresis, followed by matrix-assisted laser desorption ionization time-of-flight mass spectrometry (MALDI-TOF/MS) to identify and to verify protein differences in the early development of zebrafish (*Danio rerio*). No dramatic protein changes were observed up to 18-h postfertilization, but significant changes occurred at subsequent developmental stages. The highest number of proteins was identified 6–10-h postfertilization. From 18 to 24-h postfertilization, the number of detectable spots above 50 kDa decreased dramatically, whereas there was a surge in lower molecular weight proteins at 24-h postfertilization. Interestingly, 49% of the proteins detected at 6-h postfertilization remained detectable when fish reached an age of 1-week. Most of the identified proteins in this study were cytosolic, cytoskeletal, and nuclear proteins, and these are involved in diverse functions such as metabolism, cytoskeleton, translation, and protein degradation. Identification of such proteins also produced additional insights on what constitutes normal fish development.

A similar study was conducted on two different age groups of Atlantic cod larvae (Sveinsdottir et al. 2008), but produced very different results from those observed in the zebrafish study. Interestingly, in this second study, despite the visible morphological and functional changes that occurred in larvae from 6 to 24-days posthatch, the pattern of the most abundant proteins in these Atlantic cod larvae was largely conserved. Such differences between the two species indicate significant differences in the normal development among different fish species.

The effects of starvation in certain species, such as rainbow trout, have also been studied by using comparative protein expression tools (Martin et al. 2001). Herein, protein extracts from whole rainbow trout liver were analyzed on high-resolution two-dimensional gels. The experimental group was starved for 14-days, during which time the control animals were fed a normal daily diet until satiated. Twenty-four proteins showed abundance differences between the two groups; 8 were increased in fed fish, while 16 increased in abundance as a result of food withdrawal. Five proteins that showed a difference in abundance levels in starved and fed fish were identified, including Cathepsin D, a known lysosomal protease. This study was the first to use protein profiling in a nonmodel organism (rainbow trout) to demonstrate, for the first time in teleosts, that proteomics have the potential to assist in studying cellular mechanisms involved in protein degradation.

In another context, comparing gene and proteome expression of fish, residing in contaminated versus noncontaminated environments, may also be useful. For example, Stentiford et al. (2005) reported significant differences in fish caught from polluted and nonpolluted sites. Their study was an important pilot study conducted in 2005 on flatfish *L. limanda* in which liver lesions and normal tissue from wild populations were investigated using proteomic techniques. Liver lesions had characteristic proteomic differences, and 56 proteomic features were upregulated in tumor tissue and 20 were downregulated. Twelve of these proteomic features exhibited the potential to act as biomarkers for neoplasic lesions. These authors concluded that the proteome of cancerous liver tissue is significantly different from that of nontumor liver tissue from the same fish. This was one of the first studies, in which comparative proteomic expression was utilized as a means to discriminate between tumorous and nontumorous livers in fish. Moreover, the preliminary data from this study was one of the first to suggest that proteomic approaches may be useful in an environmental contamination context for studying fish, and may have potential application to serve as a high-throughput screening approach for disease classification.

Another interesting environmental contamination study that utilized 'omic techniques disclosed protein differences in rainbow trout that were exposed to sublethal zinc doses for several days (Hogstrand et al. 2002). Zinc exposure induced the expression of the beta-chain of the trout complement C3-1 protein. This protein plays an important role in immune response and has an immunoregulatory function (Lambris et al. 1993). The authors concluded that the induction of this protein by zinc may constitute evidence of a stimulatory effect upon trout immune system.

Ling et al. 2009 performed another study in which the effects of metal (cadmium) contamination on the fish proteome were observed in gill tissue of *Paralichthys olivaceus*, by using two-dimensional electrophoresis. Compared to a control sample, significant changes were visualized in 18 protein spots that had been exposed for 24-h to seawater contaminated with 10.0 ppm of cadmium. Among these spots, two were upregulated, one was downregulated, seven showed low expression, and eight showed high expression. Ten of the 18 proteins identified on the 2D-PAGE gel included heat shock protein 70 and calcium-binding protein, and demonstrated a synchronous response to acute cadmium toxicity. These proteins may therefore be utilized as biomarker profiles for investigating cadmium contamination levels in seawater.

Other environmental contaminants, such as cyanotoxins, have similarly been addressed using 'omics techniques. Increased attention is being played to the cyanotoxins because they bioaccumulate in fish and other aquatic species and may induce subchronic and chronic toxicity. A certain class of cyanotoxins, the mycrocistins (MC), has been recently shown to produce significant effects on the proteome expression of several fish organs (Karim et al. 2011; Mezhoud et al. 2008). Mezhoud et al. (2008) force fed adult medaka fish either an MC solution (experimental) or water (controls). After 2-h of exposure, livers were extracted and prepared for 2D-SDS-PAGE analyses. After mass spectrometric analyses, 17 differentially regulated proteins were successfully identified. The identified proteins were assigned to several functional groups, such as the following: proteins involved in cell structures, in signal transduction, enzyme regulation, and oxidative stress. Examples of the identified proteins were methyltransferase, transmembrane, apolipoprotein natural-killer-enhancing factor, and b-tubulin, among others. These altered proteins corroborated with data on studies that had mycrocistin toxic effects and modes of action, i.e., cytoskeleton disruption.

In a similar study, zebrafish were exposed to mycrocistin at two different concentrations (2 and 20 µg L^{-1}) and were then compared to a nonexposed control group. The liver proteome was also analyzed by 2D-SDS-PAGE. Compared to the 2D gels of the nonexposed zebrafish livers, the abundance of 22 protein spots from the MC-exposed zebrafish livers were significantly altered ($\geq$2-fold or $\geq$0.5-fold). Among these altered proteins, three protein spots disappeared in both groups of MC-exposed zebrafish livers, seven protein spots were significantly upregulated, and ten protein spots were noticeably down-regulated in the livers of 2 µg L^{-1} MC-exposed zebrafish. In the 20 µg L^{-1} MC-exposed group, 4 protein spots were remarkably upregulated and 11 protein spots were markedly down-regulated. The identified proteins were distinguished to comprise 22 different proteins. Of these, nine were involved in metabolism and five in proteolysis. Two proteins were characterized as cell cytoskeleton proteins, corresponding to ones that were a-actinin-4- and profilin-2-like. The other six proteins were categorized as calcium/phospholipid binding, antioxidant defense, protein folding, and other functional proteins.

One of the most interesting proteomic studies was conducted directly in the field. Rainbow trout (*Oncorhynchus mykiss*) were caged upstream and downstream from a sewage treatment works (STW), which was releasing a complex mixture of contaminants. Two-dimensional gel electrophoresis was run on liver protein extracts from these two groups, and the results showed four significantly up- and down-regulated protein spots from the group caged downstream from the STW. The three downregulated spots contained betaine aldehyde dehydrogenase, lactate dehydrogenase, and an unidentified protein, respectively. The only upregulated spot consisted of both mitochondrial ATP synthase alpha-subunit and carbonyl reductase/20b-hydroxysteroid dehydrogenase (CR/20b-HSD). This was the first study in which these types of biological responses were produced in fish exposed to a complex contaminant mixture (i.e., STW effluents). However, the significance of the differentially expressed proteins from this study is still not clear and merits further investigation.

Such techniques have primarily benefited research on well-characterized species such as humans, mice, and yeast; until now, few proteomic studies have been performed in animals from natural ecosystems (Karim et al. 2011). Unfortunately, these well-characterized species may be inappropriate from an ecotoxicological and environmental perspective, since they may not be useful as sentinel species when investigating environmental contamination (Hogstrand et al. 2002). Nevertheless, several general biomarkers that are used in mammalian models are directly transferable to fish. Karim et al. (2011) recently summarized the merge of pathways involved in environmental stress in fish and mammalian models (Fig. 1), clearly showing that several pathways are shared by both groups, i.e., the physiological process of oxidative stress. Reviews and further studies of these pathways may assist researchers to discover new biomarkers in fish and further understand existing ones.

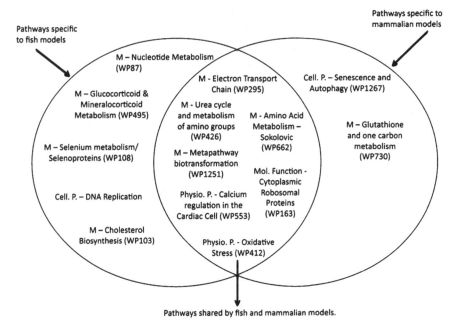

Fig. 1 Pathways involved in environmental stress in fish and mammalian models. *M* Metabolism, *Cell. P* cellular process, *Physio. P* physiological process (Adapted from Karim et al. 2011)

4 Metallomic Studies: Particularities

The recently developed field of metallomics, a term coined in the early 2000s, considers that biomolecules that bind metals and metalloids constitute a substantial portion of molecules involved in cell metabolism and behavior, and that identifying a metal cofactor in a protein can greatly assist in defining its function and placing it in the context of known cellular pathways (Haraguchi 2004). Metal-bound proteins, or metalloproteins, are now being used successfully as biomarkers of environmental exposure for some organisms (López-Barea and Gómez-Ariza 2006), although fewer studies and discoveries exist in metallomics, than for proteomics.

The late advent of the metallomic field and the paucity of metallomic findings results from several singular issues that occur when dealing with metalloprotein analyses. These issues include the absence of any protein amplification reaction similar to what exists for PCR (polymerase chain reaction) in genomics, the occurrence of post-translational changes in the biological entities, and finally, the low concentration of metallic trace elements present in biological tissues (usually <1 μg g^{-1}) and the complexity of the matrices (Gomez-Ariza et al. 2004). These deficiencies render metal-bound biomolecule analyses more difficult and challenging. Hence, metallomic analyses require sophisticated multidimensional analytical approaches. However, there has been a continuous development of new techniques that combine atomic spectroscopy and standard biochemical or proteomic techniques that enhance future prospects for metallomic analyses; these include gel electrophoresis, multidimensional capillary or nanoflow chromatography, or capillary electrophoresis, and strategies for complementary applications of element and molecule-specific detection techniques (Prange and Profrock 2005). For example, mass spectrometry approaches such as inductively coupled plasma MS (ICP-MS), electrospray MS (ESI-MS), and matrix-assisted laser desorption ionization MS (MALDI-MS) are now routinely used alongside prior 2D-PAGE protein separation to rapidly and precisely identify the metalloid component of individual proteins. Incredibly large amounts of data can be collected using these techniques, and they allow for multiend point effect screening, and the potential for identifying unexpected adverse effects (Kling et al. 2008). Unfortunately, only a few fish metalloproteins have actually been discovered and applied as biomarkers, and their functional roles in fish physiology are not yet well understood. Therefore, the challenge, and a key future goal is to use the full potential of metallomic approaches and strategies to elucidate multiple new fish metalloproteins.

5   Metalloproteins as Biomarkers of Environmental Contamination in Fish

For 5 decades, the presence of environmental contaminants in biotic and abiotic samples has been measured using chemical analytical techniques. The analyses of metal concentrations and compartmentalization in fish have also been conducted for about that long (Marigomez et al. 2002). However, since the discovery that metal speciation has toxicity implications, and more recently, with the advent of metallomics, the validity of quantifying total metal levels in fish organs, and directly relating them to environmental contamination has been questioned (Moldovan et al. 2004). Total metal levels in fish, although providing clues to toxic exposure levels, can also be influenced by the amount of metals present in the form of metal-bound proteins. So, the idea that high metal levels in certain fish organs (i.e., liver) are indicative *only* of substantial on-site contamination is no longer regarded to be true. Further insights into normal regulatory functions of essential trace-elements and metal-bound proteins in fish have contributed to this attitude change. So, while on-site contamination is still a possibility when observing high metal levels in fish, other parameters, in addition to total metal concentrations, should be analyzed. With these new ideas in effect, metalloprotein studies in aquatic organisms have begun to gain attention, and metalloproteins are now recognized as being responsive to several environmental compounds. Although the use of fish metallomics has just started, several key fish metalloproteins have been discovered and have shown utility as biomarkers for environmental contamination (Table 1). Although some proteins, e.g., several selenoproteins, still do not possess any known function, they are known to be upregulated when exposed to environmental contamination (Karim et al. 2011), and such events merit further study. Metalloprotein levels

Table 1 Studies that have addressed metalloproteins as biomarkers in an ecological context

References	Metalloprotein type	Metalloprotein name	Fish organ or compartment	Fish species	Common name	Metalloprotein role
Kimura et al. (2001)	Matrix metalloproteinases (MMPs)	MMP-2; MMP-9	Oocytes	*Oryzias latipes*	Medaka fish	Follicular development; oocyte maturation; ovulatory or postovulatory processes
Hillegass et al. (2007)		MMP13	Whole embryo	*Danio rerio*	Zebrafish	Embryogenesis
Chadzinska et al. (2008)		MMP9	Immune organs; peritoneal and peripheral blood leucocytes	*Cyprinus carpio*	Common carp	Immune response
Gladyshev (2006)	Selenoprotein	Fep 15, SelJ, SelL, SelP, SelU	Not described	Not described	Not described	Unknown
Geetha and Deshpande (1999)	Iron-binding protein	Ferritin	Liver	*Channa Punctatus, Labeo rohita, Scomberomorus cummerson, Scieluheronena Tetradactylum, Lates calcarifer*	Murrel, Rohu, Marine Spanish Mackerel, Indian Salmon, Brackish Water Perch	Important role in iron metabolism
Miguel et al. (1991) Kong et al. (2003)				*Dasyatis akajei*	Red stingray	

(continued)

Table 1 (continued)

References	Metalloprotein type	Metalloprotein name	Fish organ or compartment	Fish species	Common name	Metalloprotein role
Langston et al. (2002)	Metallothionein	–	Liver	*Anguilla anguilla*	Eel	Detoxification of heavy metals, internal homeostasis of Cu and Zn, participation in metabolic functions
Cinier et al. (1998)			Kidney, liver	*Cyprinus carpio*	Carp	
Pedrajas et al. (1993)	Superoxide dismutase	–	Liver, blood, muscle tissue	*Mugil* sp.	Mullet	Antioxidant
Li et al. (2008)	Mercury-containing protein, arsenoprotein, selenoprotein	–	Liver	*Hypophthalmichthys nobilis/ Ctenopharyngodon idella*	Bighead carp/grass carp	Detoxification processes

usually show interspecies variations and differential responses to different degrees of environmental contamination. Some of the metalloproteins that show promise as environmental biomarkers will be discussed further below.

5.1   Metallothioneins

Metallothioneins (MTs) are cysteine-rich heat-stable proteins that bind to metals entirely through metal–thiolate bonds (Kaegi and Schaeffer 1988). MTs are widely distributed, have been identified in all major classes of vertebrates, and play roles in organisms exposed to metal pollution (Roesijadi 1992; Mazzucotelli and Viarengo 1988). The mechanism of metal detoxification by MTs occurs via metal-initiated transcriptional activation of MT genes, resulting in increased MT synthesis, and subsequent binding of free metals to MT proteins. These proteins have three major physiological roles: (1) detoxification of heavy metals (Goering and Klaassen 1984); (2) internal homeostasis of Cu and Zn (Brouwer et al. 1986); and (3) participation in metabolic functions (Roesijadi et al. 1998). Several isoforms and different metals bound to these proteins exist in different fish organs (Vasak 2005).

MTs have been used successfully as biomarkers of the biological effects of heavy metals in aquatic organisms since the late 1980s (Viarengo et al. 1997). When exposed to different metals, some endogenous and exogenous factors modify MT levels in teleosts; i.e., reproductive steroids, stress hormones and season, temperature, salinity, and reproductive and dietary status (Olsson et al. 1995; Roesijadi 1992; Olsson et al. 1996; Hylland et al. 1998), and display marked variations in the inductive response among different species (Cinier et al. 1998). Therefore, stipulating what constitutes "normal" MT levels is difficult, because so many interspecies variations and differential responses to different environmental contamination levels have been observed. However, determining hepatic MT levels is an affirmed and appropriate biomarker for evaluating the biological significance of metal contamination; several studies indicate that the strength of relationships between metals and MT synthesis implies an induced response, primarily from metal exposure. Including a measurement of hepatic MT as part of a suite of sublethal effects measures, therefore, is likely to enhance environmental quality assessment (Langston et al. 2002).

MT has been often quantified, but rarely characterized, in fish tissues. The comparison between the characteristics of fish and mammalian MTs is only available in one study (Scudiero et al. 2005). What little is known indicates that fish MTs display several distinctive features in the primary structure, which includes the displacement of a cysteine residue located in the carboxy terminal half of the molecule, fewer lysine residues juxtaposed to cysteines, and the presence of three short segments of secondary structural elements. Fish MTs are also subject to heat effects, whereas mammalian forms are much less affected by heat. Moreover, fish metallothionein displays a better metal exchange capability than does mammal metallothioneines. Further 'omic studies are needed in the area of MT characterization; areas in which new work would be useful include deciphering structure, probing functional pathways, and further characterizing the nature of these metalloproteins.

5.2   Matrix Metalloproteinases

Matrix metalloproteinases (MMPs) are a family of calcium-dependent, zinc-containing endopeptidases (Bode and Maskos 2003). These proteins break down extracellular matrix components and facilitate both normal and pathological tissue remodeling, wound healing, embryo development, and tumor invasion (Matrisian 1992; Woessner 1994; Stetler-Stevenson 1996). The MMPs are important in normal physiological functions such as angiogenesis, wound healing, mammary gland and postpartum uterus involution, and cervical dilatation; however, excess MMP activity is correlated with diseases in mammalian models, such as tumor, arthritis, periodontal diseases, liver cirrhosis, atherosclerosis, and multiple sclerosis (Loftus et al. 2002; Nagase et al. 2006). Membrane-type matrix metalloproteinases (MT-MMPs) are apparently involved in cell–matrix interactions and also are implicated in a variety of immunomodulatory roles (Yang et al. 1996; Sato et al. 1997; Schonbeck et al. 1998). MMPs have also been identified in response to developmental and patho-logical changes in several organisms, primarily in the liver (Theret et al. 1998; Bueno et al. 2000). Despite the information that has been collected on the MMPs, their precise function remains unclear (Rath et al. 2001).

In fish, MMPs are not well studied, and they are only addressed in an ecological context in a few reports. MMPs were discovered in the oocytes of a small freshwater fish species, suggesting involvement during follicular development, oocyte matura-tion, or ovulatory processes (Kimura et al. 2001). These processes are directly linked to spawning and other events associated with ovulated oocytes or fertilized eggs, which are known to be affected by environmental contamination. However, the mechanism by which the MMPs affect these processes is still unknown. MMPs have also been implicated in normal and abnormal fish embryogenesis and develop-ment (Hillegass et al. 2007; Zhang et al. 2003). It has recently been demonstrated that glucocorticoid exposure alters craniofacial development and increases the expression and activity of MMPs in developing zebrafish (*Danio rerio*). Glucocorticoids are immunosuppressive and anti-inflammatory agents, used to treat autoimmune and inflammatory disorders and lymphoproliferative diseases, and are also potent teratogens (Almawi et al. 2002; Mandl et al. 2006). The mechanism by which they exert their teratogenic effects is unknown, but, in one study (Hillegass et al. 2007), zebrafish embryo glucocorticoid exposure increased the expression of two MMPs, i.e., MMP-2 (~1.5-fold) and MMP-9 (7.6–9.0-fold), at 72-h postfertil-ization. Further MMP activity was increased approximately threefold at 72-h, following glucocorticoid treatment, with craniofacial morphogenesis changes being observed. Cotreatment of zebrafish embryos with each glucocorticoid and the glucocorticoid receptor antagonist RU486 resulted in the attenuation of glucocorticoid-induced increases in MMP expression (52–84% decrease) and activity (41–94% decrease). Furthermore, the abnormal craniofacial phenotype observed following glucocorticoid exposure was less severe following RU486 cotreatment, demonstrat-ing that in embryonic zebrafish, dexamethasone and hydrocortisone alter the expres-sion and activity of MMP-2 and -9, and suggesting that these increases may be mediated through the glucocorticoid receptor.

MMPs also play a key role in fish immune responses. In one study (Chadzinska et al. 2008), sterile peritonitis was induced in common carp, when the expression of MMP-9 was evaluated in well-characterized fish immune organs, such as kidneys. The results showed that kidney MMP-9 expression was induced a mere 4-h after initial inflammation induction, and MMP-9 expression in kidney macrophages was significantly elevated until 48-h after induction. The inactivated MMP form of MMP-9, named pro-MMP-9, was also elevated up to 96-h of inflammation induction. The authors concluded that, in teleosts fish, MMP-9 should be considered as an active participant in the innate immune response, and it contributes to the resolution of the inflammation process. Although the mechanisms by which MMPs act are still unclear, they have been shown to act as useful biomarkers where inflammation processes occur.

5.3   Iron-Binding Metalloproteins: Transferrin and Ferritin

Iron-binding metalloproteins have also been discovered in fish. One main metalloprotein in this class is transferrin. It plays a crucial role in iron metabolism by binding and transporting Fe, thus making it unavailable for catalysis of superoxide radical formation. It is a single polypeptide chain of 70–80-kDa, comprised of two globular domains, resulting from an ancestral gene duplication and fusion, with each domain presenting an iron-binding site (Yang et al. 1984). Under normal conditions, most of the iron in blood plasma is bound to transferrin, and iron–transferrin complexes enter cells via a transferrin receptor-mediated endocytic pathway. Transferrin is abundant in nature and has been identified in a wide range of organisms, such as insects, crustaceans, fish, and mammals (Stafford and Belosevic 2003). Transferrin also has a close relationship with the immune system in several organisms and is recognized as a component of nonspecific humoral defense mechanisms against bacteria (Bayne and Gerwick 2001; Ellis 1999). This was first demonstrated in fish (Stafford and Belosevic 2003), wherein transferrin was expressed by activated goldfish (*Carassius auratus*) macrophages. It was also shown that transferrin significantly enhanced the killing response of goldfish macrophages exposed to different pathogens or pathogen products, such as lipopolysaccharides and several bacteria (*Mycobacterium chelonei*, *Trypanosoma danilewskyi*, *Aeromonas salmonicida*, and *Leishmania major*). This enhancement of the killing response indicated that transferrin is a primary activating molecule of macrophage antimicrobial response in fish, and is highly utilitarian as a biomarker for environmental contamination.

It has recently been discovered that transferrin is also a major cadmium-binding protein in blood plasma (De Smet et al. 2001). The authors indicate that fish transferrin shows binding affinities for cadmium in blood plasma that are comparable to human transferrin. The results of this study demonstrated that cadmium is primarily bound to two high molecular weight proteins in carp plasma, in which relatively small amounts are bound to a 60-kDa protein, and the major part is bound to transferrin. When humans and brown trout were compared to carp, differences between cadmium transport in plasma were observed. These differences were explained by the absence or very low albumin concentrations present in carp plasma, since in humans and brown trout cadmium is mainly bound to albumin, not transferrin.

Ferritin, also an iron-binding metalloprotein, plays an important role in in vivo iron metabolism (Kong et al. 2003). It is a 450-kDa protein and is the main iron storage protein in both eukaryotes and prokaryotes, and keeps iron in a soluble and nontoxic form (Chasteen 1998; Harrison and Arosio 1996). Ferritin synthesis is known to be induced when iron is available, whereas under iron deprivation conditions, ferritin synthesis is repressed (Torti and Torti 2002). Moreover, upregulation of ferritin is observed under conditions of oxidative stress (Orino et al. 2001) and inflammation (Torti and Torti 2002; Torti et al. 1988), suggesting a link to immune response. Ferritin has been described in several species by using 'omic techniques, such as northern blotting, 2D electrophoresis, molecular cloning, and nucleotide sequencing (Yamashita et al. 1996; Andersen et al. 1995; Chen et al. 2004; Geetha and Deshpande 1999; Kong et al. 2003; Miguel et al. 1991). However, ferritin analyses in fish have gained less attention. Most researchers have focused on its isolation and characterization, rather than on its functional aspects. Few ecological-related studies are available on this metalloprotein. Geetha and Deshpande (1999) used native gel electrophoresis, SDS-PAGE electrophoresis, and immunoblotting to compare ferritin characteristics among different fish species. The results indicated that the iron content of ferritins from marine and brackish species was higher than those from fresh water species, the phosphate/iron ratio was higher than mammalian ferritins, and that ferritins in fish liver are monomeric.

In another study, conducted with sea bass, both ferritin and transferrin expression in fish brain and liver were analyzed by 'omic techniques (Neves et al. 2009), after the fish were either exposed to experimental bacterial infections or to iron modulation. The fish were divided into three groups: a group receiving iron overload, another iron deficiency, and a control group. In response to infection, transferrin expression was decreased in the liver and increased in the brain. Moreover, transferrin increased in the liver, in response to iron deficiency. Ferritin expression inversely reflected transferrin content of the liver. Ferritin also increased in infection and iron overload situations and decreased in response to iron deficiency. In contrast, ferritin expression in the brain was also increased in the presence of infection.

5.4 Selenoproteins

Few selenoproteins have as yet been discovered, and even fewer have been found in fish (Gladyshev 2006). Selenium-containing proteins can be divided into three groups: proteins into which the element is incorporated nonspecifically, specific selenium-binding proteins, and specific proteins that contain selenium in the form of genetically encoded selenocysteine. There are also proteins in which selenium has been detected but for which no information on its binding form is as yet available (Behne and Kyriakopoulos 2001). Mammal selenoproteins that have known function include glutathione peroxidases, iodothyronine deiodinases, thioredoxin reductases, and selenophosphate synthetase 2. All of these are catalytically active in redox processes and are considered to be oxidoreductases that either repair or prevent damage to cellular components and regulate the redox state of proteins (Behne and Kyriakopoulos 2001; Surai 2006). Enzymatic functions for these selenium-binding proteins have been established, but information on their metabolic role and biological significance is incomplete. Mammal selenoproteins that have no known function include Selenoprotein P, Selenoprotein W, a 15-kDa selenoprotein, and an 18-kDa one.

Fish selenoproteins include the Selenoproteins U (SelU), P (SelP), J (SelJ), L (SelL), and Fep 15, the latter three of which are exclusively found in these organisms. None of these selenoproteins possess known functions, except for SelP (Gladyshev 2006). Unfortunately, little information is available on these metalloproteins.

Fep 15 is a 15-kDa selenoprotein, homologous to mammalian Sep15 and SelM, but detectable exclusively in fish, and it exhibits functions that are different from its mammalian counterparts (Novoselov et al. 2006). This selenoprotein was discovered in zebrafish, and seems to be an endoplasmatic-reticulum-resident protein that also occurs in the Golgi complex. The authors indicate that Fep 15 appears to be the first eukaryotic selenoprotein family with a highly restricted distribution. Since this protein was detected only in fish, it is suggested that Fep 15 has a specialized function unique to these organisms.

Two forms of SelP are found in fish and have a high selenium content; both appear to play a role in selenium transport and utilization (see Fig. 1) because of this high Se content (Gladyshev 2006). Very little is known about how selenium is metabolized in fish. Gladyshev (2006) suggests that the abundance of natural selenoproteins contributes to selenium accumulation in fish, and that they may have benefited from a more uniform distribution of this trace element in the earth's water reserves. SelP is also the major plasma selenoprotein, which is synthesized in the liver and delivers selenium to other organs and tissues (Gladyshev 2006; Novoselov et al. 2006).

SelU has been demonstrated to be upregulated in the medaka fish (*Oryzias latipes*), when exposed to mycrocistin, which is believed to have environmental contamination implications. Mycrocistin is a toxin produced by freshwater cyanobacteria that exists in several of the world's contaminated areas, and fish are readily exposed to these compounds through both water and diet. Even when fish were exposed to as little as 1 μg mL^{-1} of mycrocystin for 30 and 60 min durations, they still displayed significant SelU upregulation.

5.5 Hg-. Se-, and As-containing Proteins

Hg–Se proteins in blood plasma have been discovered, and they are postulated to reduce the bioavailability of toxic Hg in several organisms (Yoneda and Suzuki 1997). Recently, mercury-, arsenic-, and selenium-containing proteins have been discovered in carp livers from a mercury-polluted area in China; these proteins seem to be related to detoxification processes in carp. In the Li et al. (2008) study, three

Hg-containing bands were detected in liver of the bighead carp and one such band was detected in grass carp. The proteins present in bighead carp showed significantly higher Hg content than did those in grass carp, which may reflect different feeding habits of these species; the former ingests mainly zooplankton, whereas the latter ingests aquatic plants. This behavior may have implications for the concept of bioaccumulation of toxic trace elements, such as Hg. Proteins that bind Hg may be present in higher amounts in higher trophic-level fish that exist in contaminated areas. The authors postulated that one of the proteins that contain high Hg levels occurs in bighead carp, and may act as an Hg-storage protein. Se and As were observed to coexist in one of the Hg-containing bands taken from bighead carp and from one band from grass carp. Se has an antagonistic effect against the toxicity of Hg and As, and its presence in Hg-containing bands may reflect organismal response to environmental Hg and As contamination. Moreover, the fact that Se and As were found to coexist in the Hg-containing electrophoretic bands suggested to the authors that these two elements may be involved in a detoxification process in fish liver.

5.6 Superoxide Dismutases

Superoxide dismutases (SODs) are antioxidant metalloenzymes, classified in three major families that bind to copper or zinc, iron or manganese, or nickel. Copper/zinc superoxide dismutase (Cu/Zn–SOD) catalyzes the dismutation of superoxide to hydrogen peroxide and molecular oxygen. SODs are important enzymes in neutralizing oxygen radical-mediated toxicity. Environmental pollution may enhance oxidative stress in exposed fish, and thus disturb this natural antioxidant enzyme system (Radi and Marcovics 1988).

The effect of several contaminants, either alone or in a mixture, has been studied for the effect they have on fish SODs. For example, in one lab exposure study, in which the endocrine disruptor tri-iodothyronine (T_3) in a teleost fish (*Anabas testudines*) was evaluated, proteomic analyses were conducted by native gel electrophoresis and Western blotting (Sreejith and Oommen 2008). The fish were injected with the hormone intraperitonially daily for 5-days, after which their liver and brains were excised for proteomic analyses. Liver and brain showed a significant decrease in SOD expression after T_3 treatment. This decrease is believed to result from the oxidative stress in the fish caused by the hypermetabolic state and prolonged exposure to oxygen-free radicals.

Ken et al. (2003) studied the role of SOD in protecting organisms against a widely used herbicide, paraquat (PQ). Paraquat produces oxidative stress by generating superoxide anions and is believed to be involved in the initiation of membrane damage through lipid peroxidation. Several studies with mammalian models have demonstrated that the over-expression of SOD in cells may be associated with PQ resistance (IPCS 1984; Komada et al. 1996), but few exist in which the relationship between PQ and fish SOD have been studied. Ken et al. (2003) performed a study on 8-day-old zebrafish, in which their larvae were soaked with 0 or 55 μg ml^{-1} ZSOD for 2-h at room temperature, whereupon they were transferred to a concentration of

100-ppm PQ. Results showed that The SOD activity was increased to 1.8 times that of the untreated group. The larvae with higher SOD activity were then moved to a 100-ppm PQ solution for 24-h. The survival rate increased significantly, demonstrating that SOD does indeed protect fish against the oxidative stress of PQ toxicity.

Field studies that have involved SOD evaluations have also been conducted. Pedrajas et al. (1993) performed liver cell extracts of fish (*Mugil cephalus*) collected from polluted environments that were subject to agricultural and industrial discharges. Their work disclosed new Cu/Zn-superoxide dismutases that had high levels of Cu ions and organic compounds. These metalloenzymes have also been observed to exist in fish muscle tissue (Diaconescu et al. 2008) and in blood (Velkova-Jordanoska et al. 2008).

It has been demonstrated in laboratory studies that several proteins, including metalloproteins, are up- and/or down-regulated in response to different environmental contaminant exposures. In one such study, Mezhoud et al. (2008) reported upregulation of a selenium-binding protein and ferritin H in medaka fish liver after exposure to mycrocistin. In a similar study, Malecot et al. (2009) used this same species model and the same compound to demonstrate upregulation of transferrin. Kling et al. (2008) observed gender-specific proteomic responses in zebrafish liver following exposure to a selected mixture of brominated flame retardants, and reported the downregulation of iron regulatory protein 1 and transferrin in female fish. Wang et al. (2008) analyzed goldfish liver from fish exposed to effluent wastewater in situ, and observed up- and down-regulation of ferritin H, while Smith et al. (2009) observed that superoxide dismutase was upregulated in goldfish suffering from anoxic conditions.

6   Conclusions

The presence of environmental stressors is known to modify the proteome of organisms in specific ways. Such stressor effects have proven that proteomic studies may be quite useful as means to augment the information collected when environmental contamination studies are performed. Even in the brief time that fish protein biomarkers have been successfully used in helping to monitor for environmental contamination, they have become useful tools. Metallomics is a much newer field, and though only in its infancy, promises to become quite useful in addressing the characteristics and behaviors of metal-bound proteins, i.e., metalloproteins. Some metalloproteins are now being successfully used as biomarkers of environmental contamination, and the potential for discovering many more of these metal-bound proteins in fish is high. More research studies are needed in both field and laboratory, and these often complement each other. More work is needed in the topic areas addressed by this review, and we specifically recommend research attention to

- Identify new forms of known metalloproteins and to study their mechanisms of action.
- Search for new, previously unknown, metalloproteins.

- Further investigate whether certain metalloproteins respond to one or several types of environmental pollutants, and, if they do respond, to determine in what way.
- Conduct analyses to determine whether metalloproteins can be successfully used to address key biochemical questions when they are used as biomarkers, such as do metalloproteins vary interspecifically and/or seasonally? Does sex play a role in their expression? Does the same metalloprotein play a different role in different species? Do different maturation and life stages influence metalloprotein expression and function? Are certain biomarkers useful only when applied to some fish species, or are they applicable to fish in general?

Finally, by applying established proteomic protocols and techniques, such as 2D electrophoresis, in tandem with the new analytical techniques that are now available in the metallomic field (e.g., improved mass spectrometry and liquid chromatography), further studies in environmental contamination contexts are now possible, and will foment extensive discoveries in years to come.

7 Summary

Fish are well-recognized bioindicators of environmental contamination. Several recent proteomic studies have demonstrated the validity and value of using fish in the search and discovery of new biomarkers. Certain analytical tools, such as comparative protein expression analyses, both in field and lab exposure studies, have been used to improve the understanding of the potential for chemical pollutants to cause harmful effects. The metallomic approach is in its early stages of development, but has already shown great potential for use in ecological and environmental monitoring contexts. Besides discovering new metalloproteins that may be used as biomarkers for environmental contamination, metallomics can be used to more comprehensively elucidate existing biomarkers, which may enhance their effectiveness. Unfortunately, metallomic profiling for fish has not been explored, because only a few fish metalloproteins have thus far been discovered and studied. Of those that have, some have shown ecological importance, and are now successfully used as biomarkers of environmental contamination. These biomarkers have been shown to respond to several types of environmental contamination, such as cyanotoxins, metals, and sewage effluents, although many do not yet possess any known function. Examples of successes include MMPs, superoxide dismutases, selenoproteins, and iron-bound proteins. Unfortunately, none of these have, as yet, been extensively studied. As data are developed for them, valuable new information on their roles in fish physiology and in inducing environmental effects should become available.

Acknowledgments The authors would like to thank CNPq—National Counsel of Technological and Scientific Development for the doctorate scholarship of the main author and the Brazilian National Science and Technology Institute (Instituto Nacional de Ciência e Tecnologia—INCT de Bioanalítica) for financial support. Special thanks to David Whitacre for his invaluable suggestions, time, and patience in reviewing the manuscript.

References

Adams SM (1987) Status and use of biological indicators for evaluating the effects of stress on fish. In: Adams SM (ed) Biological indicators of stress in fish. Am Fish Soc Symp 8–18

Albertsson E, Kling P, Gunnarsson L, Larsson DGJ, Forlin L (2007) Proteomic analyses indicate induction of hepatic carbonyl reductase/20 beta-hydroxysteroid dehydrogenase B in rainbow trout exposed to sewage effluent. Ecotox Env Saf 68:33–39

Almawi WY, Abou Jaoude MM, Li XC (2002) Transcriptional and post-transcriptional mechanisms of glucocorticoid antiproliferative effects. Hematol Oncol 20:17–32

Andersen O, Dehli A, Standal H, Giskegjerde TA, Karstensen R, Rorvik KA (1995) Two ferritin subunits of Atlantic salmon (Salmo salar): cloning of the liver cDNAs and antibody preparation. Mol Mar Biol Biotechnol 4:164–170

Baker M (2005) In biomarkers we trust? Nat Biotechnol 23:297–304

Bayne BL, Brown DA, Burns K, Dixon DR, Ivanovici A, Livingstone DR, Lowe DM, Moore MN, Stebbing ARD, Widdows J (1985) The effects of stress and pollution on marine animals. Praeger, New York, p 384

Bayne CJ, Gerwick L (2001) The acute phase response and innate immunity of fish. Dev Comp Immunol 25:725–743

Behne D, Kyriakopoulos A (2001) Mammalian selenium-containing proteins. Annu Rev Nutr 21:453–473

Beyer J (1996) Fish biomarkers in marine pollution monitoring: evaluation and validation in laboratory and field studies. University of Bergen, Norway

Blom A, Harder W, Matin A (1992) Unique and overlapping pollutant stress proteins of Escherichia-Coli. Appl Environ Microbiol 58:331–334

Bode W, Maskos K (2003) Structural basis of the matrix metalloproteinases and their physiological inhibitors, the tissue inhibitors of metalloproteinases. Biol Chem 384:863–872

Bradley RW, DuQuesney C, Spargue JB (1985) Acclimation of rainbow trout, Salmo gairdneri Richardson, to zinc: kinetics and mechanism of enhanced tolerance induction. J Fish Biol 27: 367–379

Brouwer M, Whaling P, Engel DW (1986) Copper-metallothioneins in the American lobster, Homarus-americanus – potential role as Cu(I) donors to apohemocyanin. Environ Health Perspect 65:93–100

Bueno MR, Daneri A, Armendariz-Borunda J (2000) Cholestasis-induced fibrosis is reduced by interferon alpha-2a and is associated with elevated liver metalloprotease activity. J Hepatol 33:915–925

Chadzinska M, Baginski P, Kolaczkowska E, Savelkoul HFJ, Verburg-van Kemenade BML (2008) Expression profiles of matrix metalloproteinase 9 in teleost fish provide evidence for its active role in initiation and resolution of inflammation. Immunology 125:601–610

Chasteen ND (1998) Ferritin. Uptake, storage, and release of iron. Met Ions Biol Syst 35(35): 479–514

Chen SL, Xu MY, Hu SN, Li L (2004) Analysis of immune-relevant genes expressed in red sea bream (Chrysophrys major) spleen. Aquaculture 240:115–130

Cinier CD, Petit-Ramel M, Faure R, Bortolato M (1998) Cadmium accumulation and metallothionein biosynthesis in Cyprinus carpio tissues. Bull Environ Contam Toxicol 61:793–799

De Smet H, Blust R, Moens L (2001) Cadmium-binding to transferrin in the plasma of the common carp Cyprinus carpio. Comp Biochem Physiol C: Toxicol Pharmacol 128:45–53

Depledge MH, Amaral-Mendes JJ, B. Daniel RSH, Kloepper-Sams P, Moore MN Peakall DB (1992) The conceptual basis of the biomarker approach. In: (eds) D. B. Peakall L. R. Shugart Biomarkers: Research and Application in the Assessment of Environmental Health. Berlin: pp 15–29

Diaconescu C, Urdes L, Marius H, Ianitchi D, Popa D (2008) The influence of heavy metal content on superoxide dismutase and glutathione peroxidase activity in the fish meat originated frown different areas of Danube river. Romanian Biotechnol Lett 13:3859–3862

Dowling VA, Sheehan D (2006) Proteomics as a route to identification of toxicity targets in environmental toxicology. Proteomics 6:5597–5604

Ellis AE (1999) Immunity to bacteria in fish. Fish Shellfish Immunol 9:291–308

Geetha C, Deshpande V (1999) Purification and characterization of fish liver ferritins. Comp Biochem Physiol B Biochem Mol Biol 123:285–294

Gladyshev VN (2006) Selenoproteins and selenoproteomes. In: Hatfield DL, Berry MJ, Gladyshev VN (eds) Selenium: its molecular biology and role in human health. Springer Science + Business Media LLC, Philadelphia, pp 99–114

Goering PL, Klaassen CD (1984) Tolerance to cadmium-induced hepatotoxicity following cadmium pretreatment. Toxicol Appl Pharmacol 74:308–313

Gomez-Ariza JL, Garcia-Barrera T, Lorenzo F, Bernal V, Villegas MJ, Oliveira V (2004) Use of mass spectrometry techniques for the characterization of metal bound to proteins (metallomics) in biological systems. Analyt Chim Acta 524:15–22

Gonzalez-Fernandez M, Garcia-Barrera T, Jurado J, Prieto-Alamo MJ, Pueyo C, Lopez-Barea J, Gomez-Ariza JL (2008) Integrated application of transcriptomics, proteomics, and metallomics in environmental studies. Pure Appl Chem 80:2609–2626

Haraguchi H (2004) Metallomics as integrated biometal science. J Anal At Spectrom 19:5–14

Harrison PM, Arosio P (1996) Ferritins: molecular properties, iron storage function and cellular regulation. Biochim Biophys Acta – Bioenergetics 1275:161–203

Hillegass JM, Villano CM, Cooper KR, White LA (2007) Matrix metalloproteinase-13 is required for zebra fish (Danio rerio) development and is a target for glucocorticoids. Toxicol Sci 100:168–179

Hogstrand C, Balesaria S, Glover CN (2002) Application of genomics and proteomics for study of the integrated response to zinc exposure in a non-model fish species, the rainbow trout. Comp Biochem Physiol B Biochem Mol Biol 133:523–535

Hylland K, Nissen-Lie T, Christensen PG, Sandvik M (1998) Natural modulation of hepatic metallothionein and cytochrome P4501A in flounder, Platichthys flesus L. Mar Environ Res 46: 51–55

IPCS (1984) Paraquat and Diquat, environmental health criteria. WHO, Geneva, p 128

Jellum E, Thorsrud AK, Karasek FW (1983) Two-dimensional electrophoresis for determining toxicity of environmental substances. Anal Chem 55:2340–2344

Kaegi JHR, Schaeffer A (1988) Biochemistry of metallothionein. Biochemistry 27:8509–8515

Kanaya S, Ujiie Y, Hasegawa K, Sato T, Imada H, Kinouchi M, Kudo Y, Ogata T, Ohya H, Kamada H, Itamoto K, Katsura K (2000) Proteome analysis of Oncorhynchus species during embryogenesis. Electrophoresis 21:1907–1913

Karim M, Puiseux-Dao S, Edery M (2011) Toxins and stress in fish: proteomic analyses and response network. Toxicon 57:959–969

Ken C-F, Lin C-T, Shaw J-F Wu J-L (2003) Characterization of fish Cu/Zn–superoxide dismutase and its protection from oxidative stress. Mar Biotechnol 5:167–173

Kimmel DG, Bradley BP (2001) Specific protein responses in the calanoid copepod Eurytemora affinis (Poppe, 1880) to salinity and temperature variation. J Exp Mar Biol Ecol 266:135–149

Kimura A, Shinohara M, Ohkura R, Takahashi T (2001) Expression and localization of transcripts of MT5-MMP and its related MMP in the ovary of the medaka fish Oryzias latipes. Biochim Biophys Acta – Gene Struct Expr 1518:115–123

Kling P, Norman A, Andersson PL, Norrgren L, Forlin L (2008) Gender-specific proteomic responses in zebrafish liver following exposure to a selected mixture of brominated flame retardants. Ecotoxicol Environ Saf 71:319–327

Komada F, Nishiguchi K, Tanigawara Y, Akamatsu T, Wu XY, Iwakawa S, Okumura K (1996) Effect of transfection with superoxide dismutase expression plasmid on superoxide anion induced cytotoxicity in cultured rat lung cells. Biol Pharm Bull 19:274–279

Kong B, Huang HQ, Lin QM, Kim WS, Cai ZW, Cao TM, Miao H, Luo DM (2003) Purification, electrophoretic behavior, and kinetics of iron release of liver ferritin of Dasyatis akajei. J Protein Chem 22:61–70

Kultz D, Somero GN (1996) Differences in protein patterns of gill epithelial cells of the fish Gillichthys mirabilis after osmotic and thermal acclimation. J Comp Physiol B Biochem Syst Environ Physiol 166:88–100

Lambris JD, Lao Z, Pang J, Alsenz J (1993) Third component of trout complement. cDNA cloning and conservation of functional sites. J Immunol 151:6123–6134

Langston WJ, Chesman BS, Burt GR, Pope ND, McEvoy J (2002) Metallothionein in liver of eels Anguilla anguilla from the Thames Estuary: an indicator of environmental quality? Mar Environ Res 53:263–293

Li L, Wu G, Sun J, Li B, Li YF, Chen CY, Chai ZF, Iida AS, Gao YX (2008) Detection of mercury-, arsenic-, and selenium-containing proteins in fish liver from a mercury polluted area of Guizhou Province, China. J Toxicol Environ Health – Part A - Curr Iss 71:1266–1269

Ling XP, Zhu JY, Huang L, Huang HQ (2009) Proteomic changes in response to acute cadmium toxicity in gill tissue of Paralichthys olivaceus. Environ Toxicol Pharmacol 27:212–218

Loftus IM, Naylor AR, Bell PRF, Thompson MM (2002) Matrix metalloproteinases and atherosclerotic plaque instability. Br J Surg 89:680–694

López-Barea J, Gómez-Ariza JL (2006) Environmental proteomics and metallomics. Proteomics 6:S51–S62

Malecot M, Mezhoud K, Marie A, Praseuth D, Puiseux-Dao S, Edery M (2009) Proteomic study of the effects of microcystin-LR on organelle and membrane proteins in medaka fish liver. Aquat Toxicol 94:153–161

Mandl M, Ghaffari-Tabrizi N, Haas J, Nohammer G, Desoye G (2006) Differential glucocorticoid effects on proliferation and invasion of human trophoblast cell lines. Reproduction 132:159–167

Marigomez I, Soto M, Cajaraville MP, Angulo E, Giamberini L (2002) Cellular and subcellular distribution of metals in molluscs. Microsc Res Tech 56:358–392

Martin S, Cash P, Blaney S, Houlihan D (2001) Proteome analysis of rainbow trout (Oncorhynchus mykiss) liver proteins during short term starvation. Fish Physiol Biochem 24:259–270

Martin SAM, Vilhelmsson O, Medale F, Watt P, Kaushik S, Houlihan DF (2003) Proteomic sensitivity to dietary manipulations in rainbow trout. Biochim Biophys Acta – Proteins Proteomics 1651:17–29

Matrisian LM (1992) The matrix-degrading metalloproteinases. Bioessays 14:455–463

Mazzucotelli G, Viarengo A (1988) Rapid-determination of zinc, copper and cadmium organometallics in mussels by gel-permeation high-pressure liquid-chromatography and in-line detection by inductively coupled plasma atomic emission-spectrometry. Aquat Toxicol 11:416–416

Mezhoud K, Bauchet AL, Chateau-Joubert S, Praseuth D, Marie A, Francois JC, Fontaine JJ, Jaeg JP, Cravedi JP, Puiseux-Dao S, Edery M (2008) Proteomic and phosphoproteomic analysis of cellular responses in medaka fish (Oryzias latipes) following oral gavage with microcystin-LR. Toxicon 51:1431–1439

Miguel JL, Pablos MI, Agapito MT, Recio JM (1991) Isolation and characterization of ferritin from the liver of the rainbow-trout (Salmo-Gairdneri R). Biochem Cell Biol 69:735–741

Miracle AL, Ankley GT (2005) Ecotoxicogenomics: linkages between exposure and effects in assessing risks of aquatic contaminants to fish. Reprod Toxicol 19:321–326

Moldovan M, Krupp EM, Holliday AE, Donard OFX (2004) High resolution sector field ICP-MS and multicollector ICP-MS as tools for trace metal speciation in environmental studies: a review. J Anal At Spectrom 19:815–822

Nagase H, Visse R, Murphy G (2006) Structure and function of matrix metalloproteinases and TIMPs. Cardiovasc Res 69:562–573

Neves JV, Wilson JM, Rodrigues PNS (2009) Transferrin and ferritin response to bacterial infection: the role of the liver and brain in fish. Dev Comp Immunol 33:848–857

Novoselov SV, Hua D, Lobanov AV, Gladyshev VN (2006) Identification and characterization of Fep15, a new selenocysteine-containing member of the Sep15 protein family. Biochem J 394:575–579

NRC (1987) National Research Council Committee on Biological Markers – Biological markers in environmental health research. Environ Health Perspect 74:3–9

Oberemm A, Onyon L, Gundert-Remy U (2005) How can toxicogenomics inform risk assessment? Toxicol Appl Pharmacol 207:592–598

Olsson GB, Friis TJ, Jensen E, Cooper M (2007) Metabolic disorders in muscle of farmed Atlantic cod (Gadus morhua). Aquacult Res 38:1223–1227

Olsson PE, Kling P, Petterson C, Silversand C (1995) Interaction of cadmium and Estradiol-17-beta on metallothionein and vitellogenin synthesis in rainbow-trout (Oncorhynchus-Mykiss). Biochem J 307:197–203

Olsson PE, Larsson A, Haux C (1996) Influence of seasonal changes in water temperature on cadmium inducibility of hepatic and renal metallothionein in rainbow trout. Mar Environ Res 42:41–44

Orino K, Lehman L, Tsuji Y, Ayaki H, Torti S, Torti FM (2001) Ferritin and the response to oxidative stress. Biochem J 357:241–247

Pedrajas JR, Peinado J, Lopezbarea J (1993) Purification of Cu, Zn-superoxide dismutase isoenzymes from fish liver – appearance of new isoforms as a consequence of pollution. Free Radic Res Commun 19:29–41

Powers DA (1989) Fish as model systems. Science 246:352–358

Prange A, Profrock D (2005) Application of CE-ICP-MS and CE-ESI-MS in metalloproteomics: challenges, developments, and limitations. Anal Bioanal Chem 383:372–389

Quackenbush J (2001) Computational analysis of microarray data. Nat Rev Genet 2:418–427

Radi AA, Marcovics B (1988) Effects of metal ions on the antioxidant enzyme activities, protein contents and lipid peroxidation of carp tissues. Comp Biochem Physiol C 90:69–72

Rath NC, Huff WE, Huff GR, Balog JM, Xie H (2001) Matrix metalloproteinase activities of turkey (Meleagris gallopavo) bile. Comp Biochem Physiol C Toxicol Pharmacol 130:97–105

Roesijadi G (1992) Metallothioneins in metal regulation and toxicity in aquatic animals. Aquat Toxicol 22:81–114

Roesijadi G, Bogumil R, Vasak M, Kagi JHR (1998) Modulation of DNA binding of a tramtrack zinc finger peptide by the metallothionein-thionein conjugate pair. J Biol Chem 273:17425–17432

Sato H, Okada Y, Seiki M (1997) Membrane-type matrix metalloproteinase (mt-mmp) in cell invasion. Thromb Haemost 78:497–500

Schlenk D (1999) Necessity of defining biomarkers for use in ecological risk assessments. Mar Poll Bull 39:48–53

Schonbeck U, Mach F, Libby P (1998) Generation of biologically active IL-1 beta by matrix metalloproteinases: a novel caspase-1-independent pathway of IL-1 beta processing. J Immunol 161:3340–3346

Scudiero R, Temussi PA, Parisi E (2005) Fish and mammalian metallothioneins: a comparative study. Gene 345:21–26

Shepard JL, Bradley BP (2000) Protein expression signatures and lysosomal stability in Mytilus edulis exposed to graded copper concentrations. Mar Environ Res 50:457–463

Shepard JL, Olsson B, Tedengren M, Bradley BP (2000) Protein expression signatures identified in Mytilus edulis exposed to PCBs, copper and salinity stress. Mar Environ Res 50:337–340

Shi W, Chance MR (2008) Metallomics and metalloproteomics. Cell Mol Life Sci 65:3040–3048

Smith RW, Cash P, Ellefsen S, Nilsson GE (2009) Proteomic changes in the crucian carp brain during exposure to anoxia. Proteomics 9:2217–2229

Snape JR, Maund SJ, Pickford DB, Hutchinson TH (2004) Ecotoxicogenomics: the challenge of integrating genomics into aquatic and terrestrial ecotoxicology. Aquat Toxicol 67:143–154

Sreejith P, Oommen OV (2008) Tri-iodothyronine alters superoxide dismutase expression in a teleost Anabas testudineus. Indian J Biochem Biophys 45:393–398

Stafford JL, Belosevic M (2003) Transferrin and the innate immune response of fish: identification of a novel mechanism of macrophage activation. Dev Comp Immunol 27:539–554

Stegeman JJ, Brower M, Di Giulio RT, Förlin L, Fowler BA, Sanders BM, Van Veld PA (1992) Molecular responses to environmental contamination: enzyme and protein systems as indicators of chemical exposure and effect. In: RJ Huggett, Kimerle RA, Mehrle Jr PP, Bergman HL (eds) Biomarkers: biochemical, physiological and histological markers of anthropogenic stress. Lewis, Chelsea, MI, pp 235–335

Stentiford GD, Viant MR, Ward DG, Johnson PJ, Martin A, Wei WB, Cooper HJ, Lyons BP, Feist SW (2005) Liver tumors in wild flatfish: a histopathological, proteomic, and metabolomic study. Omics 9:281–299

Stetler-Stevenson WG (1996) Dynamics of matrixturnover during pathologic remodeling of the extracellular matrix. Am J Pathol 148:1345–1350

Surai PF (2006) Selenium in nutrition and health. Nottingham University Press, Nottingham

Sveinsdottir H, Vilhelmsson O, Gudmundsdottir A (2008) Proteome analysis of abundant proteins in two age groups of early Atlantic cod (Gadus morhua) larvae. Comp Biochem Physiol Part D Genomics Proteomics 3:243–250

Tay TL, Lin QS, Seow TK, Tan KH, Hew CL, Gong ZY (2006) Proteomic analysis of protein profiles during early development of the zebrafish, Danio rerio. Proteomics 6:3176–3188

Theret N, Musso O, L'Helgoualc'h A, Campion JP, Clement B (1998) Differential expression and origin of membrane-type 1 and 2 matrix metalloproteinases (mt-mmps) in association with mmp2 activation in injured human livers. Am J Pathol 153:945–954

Torti FM, Torti SV (2002) Regulation of ferritin genes and protein. Blood 99:3505–3516

Torti SV, Kwak EL, Miller SC, Miller LL, Ringold GM, Myambo KB, Young AP, Torti FM (1988) The molecular-cloning and characterization of murine ferritin heavy-chain, a tumor necrosis factor-inducible gene. J Biol Chem 263:12638–12644

Tyers M, Mann M (2003) From genomics to proteomics. Nature 422:193–197

Vasak M (2005) Advances in metallothionein structure and functions. J Trace Elem Med Biol 19:13–17

Velkova-Jordanoska L, Kostoski G, Jordanoska B (2008) Antioxidative enzymes in fish as biochemical indicators of aquatic pollution. Bulg J Agric Sci 14:235–237

Viarengo A, Ponzano E, Dondero F, Fabbri R (1997) A simple spectrophotometric method for metallothionein evaluation in marine organisms: an application to Mediterranean and Antarctic molluscs. Mar Environ Res 44:69–84

Vilhelmsson OT, Martin SAM, Medale F, Kaushik SJ, Houlihan DF (2004) Dietary plant-protein substitution affects hepatic metabolism in rainbow trout (Oncorhynchus mykiss). Br J Nutr 92:71–80

Wang JS, Wei YH, Wang DZ, Chan LL, Dai JY (2008) Proteomic study of the effects of complex environmental stresses in the livers of goldfish (Carassius auratus) that inhabit Gaobeidian Lake in Beijing, China. Ecotoxicology 17:213–220

Woessner JF (1994) The family of matrix metalloproteinases. Ann NY Acad Sci 732:11–21

Yamashita M, Ojima N, Sakamoto T (1996) Molecular cloning and cold-inducible gene expression of ferritin H subunit isoforms in rainbow trout cells. J Biol Chem 271:26908–26913

Yang FM, Lum JB, Mcgill JR, Moore CM, Naylor SL, Vanbragt PH, Baldwin WD, Bowman BH (1984) Human transferrin – Cdna characterization and chromosomal localization. Proc Natl Acad Sci USA – Biol Sci 81:2752–2756

Yang MZ, Hayashi K, Hayashi M, Fujii JT, Kurkinen M (1996) Cloning and developmental expression of a membrane-type matrix metalloproteinase from chicken. J Biol Chem 271:25548–25554

Yoneda S, Suzuki KT (1997) Equimolar Hg-Se complex binds to selenoprotein P. Biochem Biophys Res Commun 231:7–11

Zhang JS, Bai S, Tanase C, Nagase H, Sarras MP (2003) The expression of tissue inhibitor of metalloproteinase 2 (TIMP-2) is required for normal development of zebrafish embryos. Dev Gene Evol 213:382–389

Spatial Distribution of Arsenic in Groundwater of Southern Nepal

Ishwar Chandra Yadav, Surendra Singh, Ningombam Linthoingambi Devi,
Devendra Mohan, Madhav Pahari, Pratap Singh Tater,
and Birendra Man Shakya

Contents

I.C. Yadav • S. Singh (✉)
Centre of Advanced Study in Botany, Banaras Hindu University,
Varanasi 221005, UP, India
e-mail: surendrasingh.bhu@gmail.com

N.L. Devi
Key Laboratory of Biogeology and Environmental Geology, China University of Geosciences,
388, Lumo Road, Wuhan 430074, People's Republic of China

D. Mohan
Department of Civil Engineering, Institute of Technology, Banaras Hindu University,
Varanasi 221005, UP, India

M. Pahari
Water, Sanitation and Hygiene Specialist, UNICEF Country Office,
Pulchowk, Kathmandu, Nepal

P.S. Tater
CHHINMASTA Consultancy Private Limited, Putali Sadak, Kathmandu, Nepal

B.M. Shakya
Water Quality Improvement and Monitoring Section, Department of Water Supply and Sewerage,
Panipokhari, Kathmandu, Nepal

D.M. Whitacre (ed.), *Reviews of Environmental Contamination and Toxicology,*
Reviews of Environmental Contamination and Toxicology 218,
DOI 10.1007/978-1-4614-3137-4_3, © Springer Science+Business Media, LLC 2012

1   Introduction

Groundwater is a significant source of drinking water in virtually all parts of the world. Protected groundwater is safer to drink, in terms of microbiological quality, than is water from open dug wells and ponds (World Bank 2005). However, groundwater is notoriously prone to chemical and other types of contamination, such as arsenic (As) that derives from natural or anthropogenic sources.

Groundwater is abundant in the Quaternary alluvial sediments of the lowland Terai region of southern Nepal and is the only source of drinking water for the people in the Terai region (CBS 2004; Kansakar 2006). This region is estimated to have approximately 1,000,000 tubewells, which supply groundwater for some 13 million people (UNICEF 2006). Both shallow and deep aquifers exist throughout most of the Terai region (Jacobson 1996). However, mostly shallow groundwater aquifers are used to supply drinking water. These shallow aquifers appear to be primarily unconfined and permeable, although they are sparse or absent in some areas (Upadhyay 1993). Shallow aquifers are more susceptible to As contamination, whereas deep aquifers are generally free of As contamination, although the latter are more expensive to tap (Bisht et al. 2004). Deep tubewells may also become contaminated if the practices used to construct them are improper. Geology and geomorphology are the main factors that control the magnitude and incidence of As that appears in the groundwater of the Terai region (Kansakar 2004). The southern districts of this region are characterized by having a thick clay layer of low permeability over basal gravel (Dixit and Upadhya 2005; JICA 1990).

Several studies have been conducted on the quality of groundwater in the Terai region, and these have revealed the presence of high concentrations of As in shallow tubewells (<50-m depth), although most analyzed samples display As concentrations that are less than 10 μg L^{-1} (Chaturvedi 2003; Dahal et al. 2008; Gurung et al. 2005; Mahat and Kharel 2009; Mahat and Shrestha 2008; Panthi et al. 2006; Sharma et al. 2005; Shrestha et al. 2003; World Bank 2005; Yadav et al. 2011). In this review, we have summarized the literature that exists on the extent of As contamination in the six Terai districts (Sunsari, Dhanusha, Bara, Rupandehi, Kailali, and Kanchanpur) of Nepal. In so doing, we intend to provide a perspective of how As contamination of drinking water is distributed in this part of Nepal.

2   Monitoring for As Residues in Nepal

By early in the year 2000, several Nepalese governmental and some nongovernmental bodies realized that there was a need to address the serious health implications of the As poisoning that resulted mainly from certain Nepalese populations consuming As contaminated ground water; hence, this cooperative initiated testing for As levels in tubewells, and began a mitigation program. In 2001, monitoring of 4,000 tubewells that were distributed throughout the 20 Terai districts was initiated by the

Department of Water Supply and Sewerage (DWSS)/UNICEF, and this was the first major investigation of As levels in groundwater. From this study, it was discovered that more than 10% of the tested tubewells had water that exceeded the WHO guideline of 10 μg L^{-1}, whereas water in 3% of these tubewells exceeded the Nepalese interim standard (NIS) guideline (50 μg L^{-1}) for As content. In 2003, the National Arsenic Steering Committee (NASC) was formed. The NASC is a national level steering committee designed to address As contamination, and was constituted to coordinate and streamline all As activities, such as maintaining uniformity, and providing policy-level guidance, developing standards, performing testing, facilitating mitigation, and communication activities with the concerned ministry. The NASC worked in collaboration with the Environment Public Health Organization (ENPHO) to perform testing on 18,635 tubewells in 20 Terai districts, under a program called the "State of Arsenic in Nepal 2003." The results of this study showed that 23.7 and 7.4% of the tubewells retained As level that exceeded the WHO and NIS guidelines, respectively (NASC/ENPHO 2003). Moreover, after 2004, the DWSS and Nepal Red Cross Society (NRCS), with financial assistance from UNICEF, initiated a blanket testing program in all 20 Terai districts. By March 2007, nearly 640,000 tubewells distributed throughout the 13 districts of the region had been tested. Of these 13 districts, eight (viz., Kanchanpur, Nawalparasi, Rautahat, Kapilbastu, Parsa, Siraha, Sarlahi, and Saptari) were included under the program "The State of Arsenic in Nepal-2005," testing that was performed by NASC/UNICEF that took place during 2005–2006. From such testing blanket, testing data were compiled that represented 327,641 tubewells, plus the additional 11,308 tubewells associated with the "State of Arsenic in Nepal 2003" study.

The results from the abovementioned studies showed that 13 and 2.8% of total tested tubewells had As levels in water that exceeded the WHO and NIS guideline values (10 and 50 μg L^{-1}), respectively (NASC/UNICEF 2007). In 2005, under a blanket testing program of the "The State of Arsenic in Nepal 2005" study, a national arsenic database was conceived and assembled as an integrated information management system tool. This assemblage was called the "Arsenic Information Management System (AIMS)," and was developed to standardize the common database format so that it could enable data consistency and interoperability. This effort helped facilitate the sharing and proper dissemination of As data between various stakeholder organizations. The main features of the AIMS project constituted an effort to manage the As database and associated GIS interfaces, to enhance analysis and reporting of data, and to provide valuable information needed for prioritizing, planning, and implementing As mitigation measures. Five districts namely, Sunsari, Dhanusha, Bara, Rupandehi, and Kailali and nine village development committees (VDC: the lower administrative division of district) of the Kanchanpur district were included in the studies conducted by NASC/UNICEF (NASC/UNICEF 2007). Hach field kits are a useful tool for assessing As concentrations in water in the field; test results are based on the reduction of inorganic arsenic to arsine gas (AsH$_3$), which is allowed to pass through the mercury bromide (HgBr$_2$) indicator paper and the intensity of color indicates the concentration of arsenic. Such kits were used for testing As throughout the blanket test districts. The reliability of Hach field kits was cross checked with laboratory tests using atomic absorption spectrophotometry (AAS). In this chapter, we have processed and analyzed the data from six southern districts of the region based on the blanket testing results completed from As analyses in Sunsari, Dhanusha, Bara, Rupandehi, Kailali, and Kanchanpur (NASC/UNICEF 2007).

3 As Contamination of Groundwater in Nepal

3.1 Geographic Distribution of Arsenic Concentrations

The geographic distribution of the tubewells tested for As was not uniform in the Terai districts of Nepal. Among the six Terai districts, tubewells were most frequently tested in Kailali (84,205 have been tested), followed by Rupandehi (72,703), Sunsari (66,664), Dhanusha (56,531), Kanchanpur (53,578), and Bara (38,589) (Table 1). The total number of tubewells tested, and the levels of As identified in the six districts of Terai, is shown in Fig. 1.

The As levels found in water from tubewells in Terai ranged from 0 to 770 $\mu g\,L^{-1}$, with majority (93%) of the tubewells containing As concentrations below 10 $\mu g\,L^{-1}$ (Fig. 2). Approximately 5% of the tested tubewells contained water with As levels that fell between 11 and 50 $\mu g\,L^{-1}$, and the remaining 2% had As concentrations exceeding 50 $\mu g\,L^{-1}$. Thus, 7% of the tested tubewells had As concentrations above the WHO limit, whereas only 2% of these tubewells had As levels exceeding the NIS guideline. The maximum concentration of As found (770 $\mu g\,L^{-1}$) occurred in Khomada (of Dhansinhapur VDC) in the Kailali district. The next highest residues found, in descending order, were in Nauwakhor Prashahi, Paterwa VDC (500 $\mu g\,L^{-1}$) in Dhanusha, Baisi Bichawa VDC (500 $\mu g\,L^{-1}$) in Kanchanpur, and Bagaha VDC (500 $\mu g\,L^{-1}$) in the Rupandehi districts. Further, Bhatauda VDC of Bara district and Madhyeharsahi VDC in Sunsari district had As levels of 350 and 300 $\mu g\,L^{-1}$, respectively, in their tested tubewells.

Table 1 Distribution of arsenic concentration levels in tubewells of the various Nepalese districts of Terai region

District	0–10 ppb	11–50 ppb	>50 ppb	Total
Sunsari	63,903	2,343	418	66,664
Dhanusha	54,388	1,724	419	56,531
Bara	34,444	2,689	1,456	38,589
Rupandehi	69,950	2,283	470	72,703
Kailali	74,357	7,009	2,839	84,205
Kanchanpur	47,633	4,365	1,580	53,578
Total	344,675	20,413	7,182	372,270

Source: NASC/UNICEF (2007)

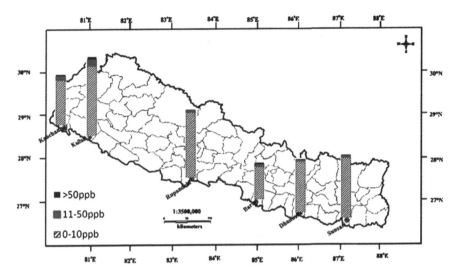

Fig. 1 The distribution of arsenic contamination in tested tubewells of the six Terai districts in Nepal

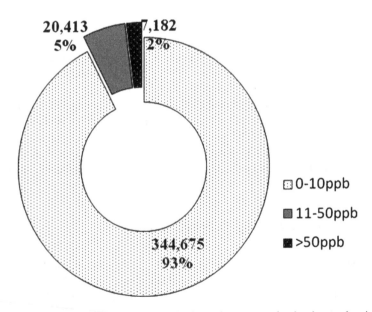

Fig. 2 The classification (percent distribution) of arsenic concentration levels prevalent in tested tubewells of the six districts of the Terai region

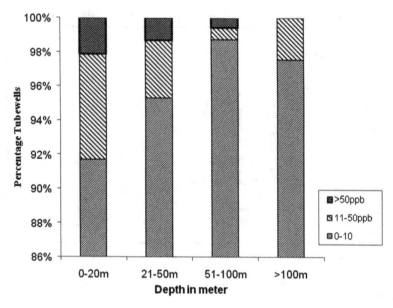

Fig. 3 The distribution, by tubewell depth, of As levels measured in well waters collected in Sunsari, Dhanusha, Bara, Rupandehi, Kailali, and Kanchanpur districts

The mean As concentration varied from district to district, with Kailali having the highest mean As concentration (6.27 µg L^{-1}), followed by Kanchanpur (4.98 µg L^{-1}) and the Bara (4.94 µg L^{-1}) districts. The mean As concentration in the other three districts was below 2 µg L^{-1}. Using the NIS guideline as a reference point, the As levels detected in the three districts of the six that had the highest residues, expressed as a percentage of the NIS guideline, were as follows: the Bara district (highest; 3.8%), Kailali (3.4%), and Kanchanpur (2.9%). However, if similar values are based on the WHO standard then As levels in Kailali had the highest percentage of tubewells (11.7%) that exceeded the standard, followed by Kanchanpur (11.1%), Bara (10.7%), Sunsari (4.1%), and Dhanusha and Rupandehi (3.8%).

3.2 Distribution of As by Tubewell Depth

Three types of tubewells are used as a source of drinking water in the Terai region, and they vary by depth. These three are shallow tubewells (STW) (<50-m deep), deep tubewells (DTW) (>50-m deep), and dug wells (DW) (up to 20-m or more deep). Out of the total of the tested tubewells, a majority (98%) were STWs. However, 0.3 and 0.8% of the tested tubewells were DTWs or DW, respectively. The depth of DWs in the tested districts displayed various As concentrations. The depth of these DWs, ranged from <1 to 137-m, and they had a mean and median depth of 16 and 10.6-m, respectively (Fig. 3). The depth of deep tubewells ranged from 1 to 183-m, and had mean and median depths of 47.7 and 45.7-m, respectively.

Virtually all (97%) of the tested tubewells that had As levels exceeding both WHO and the NIS guidelines, had a depth of <20-m. At this depth range, more than 8% of tubewells had As levels above 10 μg L^{-1}, while only 2% of tubewells had levels above 50 μg L^{-1}. At a depth of 21–50-m, 4.7 and 1.3% of the water in tubewells had As concentrations that exceeded the 10 and 50 μgL^{-1} guideline levels, respectively. Similarly, at a depth greater than 50 m, tubewells having an As concentration that exceeded guideline values (10 and 50 μgL^{-1}) were significantly fewer in number. Therefore, it seems that tubewells having a depth less than 20-m had average higher As concentrations. In contrast, the deeper tubewells contained lower average As concentrations. However, about 3.5% of shallow tubewells that had depths greater than 100-m showed As levels that exceeded 10 μg L^{-1}. Similarly, 0.6 and 4% of deep tubewells that had depths between 51 and 100-m contained As levels above 10 and 50 μg L^{-1}, respectively. About 4 and 0.58% of the deep tubewells having a depth of <20-m contained As level above the 10 and 50 μg L^{-1} guideline levels, respectively. About 3.2 and 1% of the tested dug wells having depths up to 20-m contained As levels above 10 and 50 μg L^{-1}, respectively. Approximately 1.5 and 0.4% of dug wells that were 21–50-m deep contained As level above 10 and 50 μg L^{-1}, respectively. Only 2.2% of the dug wells having a depth above 50-m contained As levels above 10 μg L^{-1}.

The correlation of As concentration with depth of shallow tubewells were statistically significant ($p < 0.0001$), but had a positively weak correlation ($\rho = 0.0464$) (Fig. 4 and Table 2). This relationship confirms that an increase in As levels existed in shallow tubewells, particularly at depths <50-m. In contrast, the level of As decreased in tubewells that were deeper than 70-m. This infers that shallow tubewells having depths up to 50-m showed higher levels of As. Similarly, the correlation between As concentration and depth of deep tubewells showed a negatively weak correlation ($\rho = -0.1736$). When these data are placed on a scatter plot, they also show that a significant number of deep tubewells have higher arsenic levels up to 50-m depth (Fig. 4). The As level tends to decrease at depths below 50-m, as shown by the decreasing slope of the regression line. Hence, it can be inferred that deep tubewells having depths below 50-m are not excessively contaminated by As, and the water therein can be regarded as being safe. Similarly, the As concentration was statistically significant ($p = 0.0024$), but had a negatively weak correlation ($\rho = -0.057$) as depth of dug wells increased. The majority of dug wells had higher As levels, up to a depth of 20-m. The As levels tended to decrease below a depth of 20-m, as shown by the decreasing slope of the regression line (Fig. 4). Therefore, it seems that dug wells having depths greater than 20-m are safe from an As contamination standpoint.

3.3 Classification of Arsenic Concentration

There are a total of 408 VDCs/municipalities in the six Terai districts (CBS 2004), of which only 373 VDCs have had their tubewells tested for As. Among those areas that

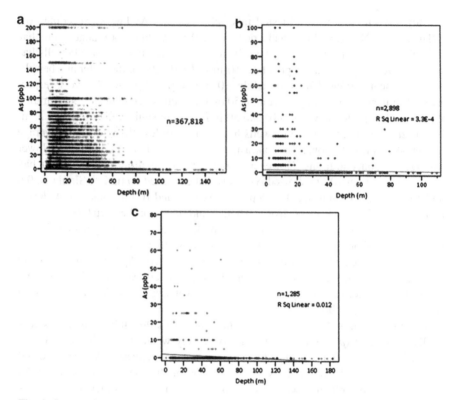

Fig. 4 Scatter plot graphs of arsenic concentration levels by depth in: (**a**) shallow tubewells (**b**) deep tubewells, and (**c**) dug wells

Table 2 Paired sample Spearman correlation of arsenic level with depth of tubewells

Types of tubewells	No. of tubewells (N)	Correlation (r)	Significance (p)
Shallow tubewells	367,818	0.0464	<0.0001
Deep tubewells	2,898	−0.1736	<0.001
Dug wells	1,285	−0.057	0.0024

have had tubewells tested for As contamination, the wells have been divided into three classes (1) wells that contain As levels less than the WHO guideline (0–10 µg L^{-1}), (2) wells with As residues laying between the WHO and NIS guidelines (11–50 µg L^{-1}), and (3) wells having As residues that exceeded the NIS guideline (>50 µg L^{-1}). The frequency of As detections, among the six districts, is shown on a map presented in Fig. 5. This map illustrates the distribution of As levels among the regions, as presented in three concentration ranges: 0–10 µg L^{-1}, 11–50 µg L^{-1}, and above 50 µg L^{-1}. Although the Kailali, Rupandehi, and Sunsari districts had the largest number of tubewells tested, the Bara, Kailali, and Kanchanpur districts had a higher proportion of tubewells that displayed As levels exceeding 50 µg L^{-1}.

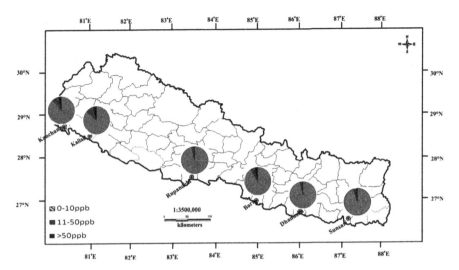

Fig. 5 The distribution of As levels in tested tubewells to which the populations of the six districts of southern Nepal were exposed

Tubewells that had water with As residues exceeding the NIS and WHO guidelines, or residues that only exceeded the NIS guideline, were more frequently found in the Bara and Kailali districts. However, tubewells having As concentrations above the WHO guideline were more frequently found in the Rupandehi and Dhanusha districts. The number of tubewells, having As concentrations between 11 and 50 μg L^{-1}, were most frequently found in the Kailali district. However, none of the districts had tubewells with As concentrations below the WHO guideline.

3.4   Human As Exposure

A population of nearly 5.25 million people, representing 542,064 households in the six Terai districts of Nepal, uses wells that contain As contamination for sourcing their drinking water. About 406,430 (7.7%) people, representing 37,443 (6.9%) households, are exposed to As concentrations that exceed the WHO guideline, and a population of 117,284 (2.2%) are compelled to drink water containing As level above the NIS guideline (Table 3). The highest percentage of population that is exposed to As contamination is in the Kailali district, wherein the majority (4%) of the population (>43,000 people) relies on contaminated tubewell water that commonly has As levels above 50 μg L^{-1}. The next highest As residue exposure takes place in the Bara district, wherein 3.6% (over 34,000 people) of the population are at risk of significant As exposure to levels exceeding the NIS guideline value. Similarly, 3.5% of the total population (>20,500 people) in the Kanchanpur district

Table 3 The number of individuals from different Nepalese districts that have exposure to different As-contamination levels

District	0–10 ppb	11–50 ppb	>50 ppb	Total
Sunsari	824,621	34,826	8,200	867,647
Dhanusha	783,632	26,044	4,991	814,667
Bara	873,600	51,839	34,236	959,675
Rupandehi	974,044	30,334	6,172	1,010,550
Kailali	878,063	98,816	43,166	1,020,045
Kanchanpur	510,688	47,287	20,519	578,494
Total	4,844,648	289,146	117,284	5,251,078

Source: NASC/UNICEF (2007)

have no reasonable alternative but to use As-contaminated tubewell water, which contains As levels exceeding 50 µg L^{-1}.

3.5 Exposure Vulnerability via Drinking Water

The VDCs was assessed with the relative vulnerability potential for As contamination in tubewell drinking water, and have discovered that it is very high. This high exposure consequently increases the risk of As-related health hazards that result from consumption of As contaminated water. Neither the distribution of As-contaminated drinking water, nor the distribution of As-tested tubewells is homogeneous at the district/VDC level. Therefore, all As-tested districts/VDCs cannot and do not have equal vulnerability to As exposure or potential health effects. Some districts/VDCs are more vulnerable to As intake than in others. Therefore, a basic criterion was developed by National Arsenic Steering Committee (NASC) to classify the level of geographical vulnerability that exists to As. The classification was based on the distribution of the As-tested tubewells, the degree that As residues in them exceeded WHO and NIS guidelines; four vulnerability classes were defined, as follows: the VDCs/municipalities are categorized as being either low vulnerable, moderately vulnerable, moderately high vulnerable, or highly vulnerable to As exposure. The VDCs/municipalities, where the percentage of tubewells having As concentration exceeding 10 µg L^{-1} were zero, are categorized as being "low vulnerable." The VDCs/municipalities for which the percentage of tubewells having As concentrations exceeding 10 µg L^{-1} ranged between 1 and 25 are categorized as being "moderately vulnerable." VDCs/municipalities for which the percentage of tubewells having As concentrations between 26 and 50, or above 50 are categorized as being "moderately high vulnerable" or "highly vulnerable," respectively.

The country map of drinking water vulnerability to As also presents the status of As levels in six Terai district. The vulnerability classes of As exposure for each Terai district is presented as one of four levels of vulnerability, as shown in Fig. 6. Kailali is the only district which was labeled as having a "moderately high vulnerability" to As exposure. The reason for this rather high vulnerability was that 50%

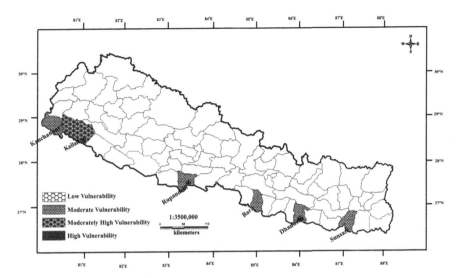

Fig. 6 Map showing the relative degree of vulnerability to drinking water in different districts of Nepal

Table 4 The relative vulnerabilities of various Nepalese VDC and Municipalities to arsenic contamination and exposure

District	No. of VDC tested	Low vulnerable	Moderate vulnerable	Moderately high vulnerable	High vulnerable
Sunsari	49	11	37	1	–
Dhanusha	99	33	62	4	–
Bara	97	22	62	10	3
Rupandehi	71	13	58	–	–
Kailali	37	3	29	4	1
Kanchanpur	20	1	16	2	1
Total	373	83	264	21	5

Sources: NASC/UNICEF (2007); *VDC* village development committee

of the tested tubewells of the district had As levels that were above WHO guidelines. Of the other five districts (viz., Sunsari, Dhanusha, Bara, Rupandehi, and Kanchanpur) all are classified as "moderately vulnerable" to As exposure. Not even one district was found to be either "highly vulnerable" to As exposure, or was found to have a "low vulnerable" classification to As exposure.

There were 373 VDC/municipalities that were blanket tested for As vulnerability. Of these, six districts, which constituted a majority (71%) rated as being "moderately vulnerable," 22% were classified as being "low vulnerable," 5.6% as "moderately high vulnerable," and 1.3% as "highly vulnerable." The five "highly vulnerable" VDCs were as follows: Dharmanagar, Matiarwa, and Piparpati Parchrouwa in the Bara district; Kota Tulsipur in the Kailali district; and Laxmipur in the Kanchanpur district (Table 4). Among the "highly vulnerable" VDCs,

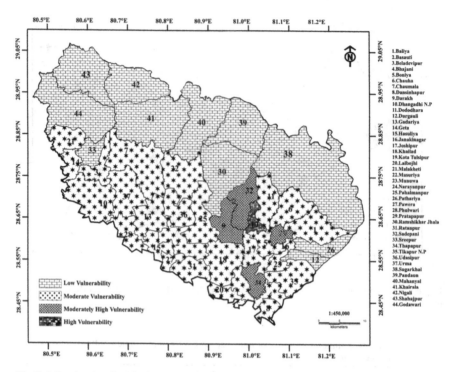

Fig. 7 Map showing degree of vulnerability to As in the groundwater of different VDCs of Kailali district

Dharmanagar in Bara district stands first with 60% of tubewells exceeding an As level of 10 μg L⁻¹, followed by Kota Tulsipur (58%) in the Kailali district. In Fig. 7, a map of the Kailali district is presented, which shows the "high vulnerability" results of As exposure.

3.6   Arsenic Hotspots in the VDCs

Hotspots were defined as areas that had very high (>50 μg L⁻¹) As contamination of drinking water. In total 7,182 tubewells, from 204 VDCs/municipalities in six districts had As concentrations that exceeded the NIS limit of 50 μg L⁻¹. In the majority of these VDCs, only a few tubewells contained an As concentration higher than 50 μg L⁻¹. The communities/localities located and mapped those tubewells in their areas that had As levels exceeding the 50 μg L⁻¹, and these were noted on their district hotspot maps. Kota Tulsipur VDC of the Kailali district had the highest number of tested tubewells that showed As concentrations above 50 μg L⁻¹. A map of Kota Tulsipur that displays the distribution of the different degrees of As contamination is presented in Fig. 8.

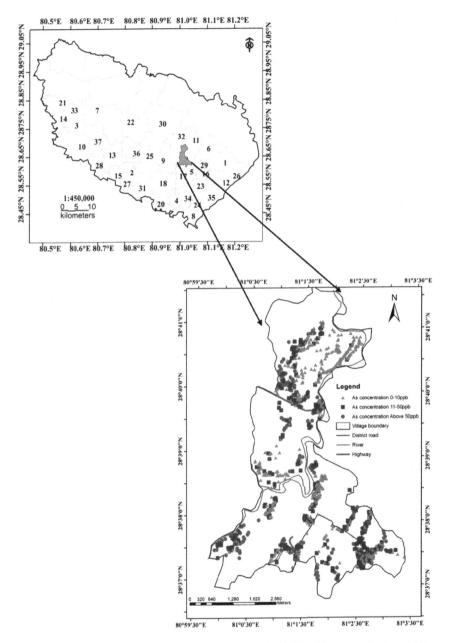

Fig. 8 The distribution of arsenic by residue levels in the highly affected VDC Kota Tulsipur of Kailali district

4   Conclusions

Multiple monitoring studies have been conducted by governmental and other bodies of As contamination in several districts of Nepal. In total, the water from 372,270 tubewells have been monitored for As contamination. From the review of these data, we conclude the following:

1. The existing database on As contamination in Nepalese tubewell and dug wells has grown to extensive proportions since 2000. However, several key additional steps are needed to enhance the work performed so far. Among these are the following:

    (a) Geological and hydrogeological investigations to identify distribution patterns throughout the district.
    (b) Construction of a database system of As contaminated areas and tubewells.
    (c) Studies to determine the exact number of people drinking contaminated water above 50 μg L^{-1} to assess magnitude of the calamity.
    (d) Perform a survey of health and mitigation activities to better determine the number of people clinically and subclinically affected by As contamination.
    (e) Identify which aquifer around As-hotspots areas is safe.
    (f) Adopt a mitigation strategy to prevent further As-exposures.

2. The results of As monitoring to date shows the following:

    (a) As levels are below critical safe/guidelines values in approximately 93% of all samples analyzed.
    (b) Although sampling sites are not uniformly distributed, which could cause statistical distortions, the highest As levels were found in Khomada village of Kailali district.
    (c) As residues varied by depth with higher residues occurring at depths between 0 and 20-m.
    (d) There is gradual increase in As level in shallow tubewells up to a depth of 50-m, which tends to decrease at depths below 70-m.
    (e) Human exposure occurs to As and was most prevalent in Kailai districts, followed by the Bara district.
    (f) Depending upon the level of geographical vulnerability to As, the Kailali district was found to be more vulnerable.
    (g) The Kota Tulsipur VDC of Kailali district was identified as a As hotspot VDC among 204 tested VDC/Municipalities.

5   Summary

Groundwater is an important source of both drinking water and of irrigation in the Terai region of Nepal. Although thousands of tubewells have been drilled in this region, the distribution of those that have been sampled and analyzed for As contamination is not consistent across the region. Based on a recent blanket tubewell testing program conducted in Nepal in 2007, preliminary data are available that allows us to provide a perspective on the As-contamination situation in drinking water of the six southern the Terai districts of Nepal. Arsenic concentrations detected in drinking water of tubewells and dug wells in these districts ranged from 0 to 770 μg L^{-1}. It was found that the majority of the tested wells contained water that had As level below 10 μg L^{-1}, which is the WHO guideline value for exposure to As. The mean As levels detected varied from 2 μg L^{-1} in the Sunsari, Dhanusha, and Rupandehi districts, to 6.27 μg L^{-1} in the Kailali district. The distribution of As levels detected, based on the NIS guideline, and expressed as a percentage thereof, was highest in the tubewells of the Bara district (3.8%), followed by the Kailali district (3.4%). Wells that were between 0- and 20-m deep contained water that had the highest percentage of As residues that exceeded both the WHO and NIS guideline values. In shallow tubewells of all six tested districts, the highest As contamination levels were found at depths of <50-m. The proportion of the population that was most often exposed to As levels >50 μg L^{-1} occurred in the Kailali district (4%), followed by the Bara district (3.6%). Using a system developed to classify the relative vulnerability of inhabitants to As through drinking water only the Kailali and Bara districts were classified as "highly vulnerable" to As exposure. The Kota Tulsipur VDC of the Kailali district was found to be the most prominent As hotspot, wherein the majority of tubewells contained more than 50 μg L^{-1} of As.

Acknowledgements ICY is thankful to Indian Council for Cultural Relation (ICCR) for financial assistance in the form of South Asian Association for Regional Cooperation (SAARC) fellowship (F.N.8-2/09 10/SAARC/ISD-II). Thanks are also due to Genesis Consultancy, Kathmandu, and UNICEF Nepal, Kathmandu, for providing relevant data.

References

Bisht SB, Khadka MS, Kansakar DR, Tuinhof A (2004) Study of arsenic contamination in irrigation tubewells in the Terai, Nepal. In: Kansakar DR (ed) Proceeding of the Seminar on Arsenic Study in Groundwater of Terai & Summary Project Report, Arsenic Testing and Finalization of Groundwater Legislation Project, 2003, 28 December, Kathmandu, Nepal. pp 31–47

CBS (2004) Environment statistics of Nepal. Central Bureau of Statistics, Kathmandu, Nepal, Kathmandu, Nepal

Chaturvedi CP (2003) Arsenic testing and finalization of ground water legislation project. Department of Irrigation/Nepal Government, pp 1–15

Dahal BM, Fuerhackerb M, Mentlera A, Shrestha RR, Bluma WEH (2008) Screening of arsenic in irrigation water used for vegetable production in Nepal. Arch Agron Soil Sci 54:41–51

Dixit A, Upadhya M (2005) Augmenting groundwater in Kathmandu valley: challenges and possibilities, Nepal Water Conservation Foundation 2005 http://www.iah.org/recharge/downloads/NepalpaperJan05final.pdf. (Accessed on 26 June 2008)

Gurung JK, Hiroaki I, Khadka MS (2005) Geological and geochemical examination of arsenic contamination in groundwater in the Holocene Terai Basin, Nepal. Environ Geol 49:98–113

Jacobson G (1996) Urban groundwater database. AGSO Report, Australia 1996, http://www.clw.csiro.au/UGD/DB/Kathmandu/Kathmandu.html

JICA (1990) Groundwater management project in the Kathmandu Valley. Final Report to Nepal water supply cooperation 1990. Japan International Cooperation Agency

Kansakar DR (2006) Understanding groundwater for proper utilization and management in Nepal. In: Sharma BR, Villholth KG, Sharma KD (eds) Groundwater research and management: integrating science into management decisions. In: Proceedings of IWMI-ITP-NIH International Workshop on "Creating Synergy between Groundwater Research and Management in South and Southeast Asia, 2005", 8–9 February, Roorkee, India

Kansakar DR (2004) Geologic and geomorphologic characteristics of arsenic contaminated groundwater areas in Terai, Nepal. In: Kansakar DR (ed) Proceeding of the seminar on arsenic study in groundwater of terai and summary project report, arsenic testing and finalization of groundwater legislation project. pp 31–47

Mahat RK, Shrestha R (2008) Metal contamination in ground water of Dang district. Nepal J Sci Technol 9:143–148

Mahat RK, Kharel RP (2009) Status of arsenic contamination and assessment of other probable heavy metal contamination in groundwater of Dang distrcit in Nepal. Scientific World 7:33–36

NASC/ENPHO (2003) The state of arsenic in Nepal-2003, Kathmandu, Nepal: National Arsenic Steering Committee/Environment Public Health Organization

NASC/UNICEF (2007) Report on blanket tube well testing in sunsari, Bara, Dhanusha, Rupandehi, Kailali and Kanchanpur Districts 2007. A report prepared for NASC/UNICEF by Genesis Consultancy (P) Ltd., Kathmandu

Panthi SR, Sharma S, Mishra AK (2006) Recent status of arsenic contamination in groundwater of Nepal-a review. Kathmandu Univ J Sci Eng Technol 2:1–11

Sharma S, Bajracharya RM, Sitaula BK, Merz J (2005) Water quality in the Central Himalaya. Curr Sci 89:774–786

Shrestha RR, Shrestha MP, Upadhyay NP, Pradhan R, Khadka R, Maskey A, Maharjan M, Tuladhar S, Dahal BM, Shrestha K (2003) Groundwater arsenic contamination, its health impact and mitigation program in Nepal. J Environ Sci Health Part A 38:185–200

UNICEF (2006) Diluting the pain of arsenic poisoning in Nepal. Retrieved 11 12, 2007, from UNICEF Nepal-diluting the pain of arsenic poisoning in Nepal: http//www.unicef.org/inforby-country/nepal_35975.html

Upadhyay SK (1993) Use of groundwater resources to alleviate poverty in Nepal: policy issues. In: Kahnert F, Levine G (eds) Groundwater irrigation and the rural poor: options for development in the Gangetic Basin. World Bank, Washington, DC

World Bank (2005) Towards more effective operational response: arsenic contamination of groundwater in South and East Asian countries. Volume II – Technical Report, water and sanitation program, South Asia Region

Yadav IC, Dhuldhaj UP, Mohan D, Singh S (2011) Current status of groundwater arsenic and its impacts on health and mitigation measures in the Terai basin of Nepal: an overview. Environ Rev 19:56–69

Propoxur: A Novel Mechanism for Insecticidal Action and Toxicity

Peter Kovacic and Ratnasamy Somanathan

Contents

1 Introduction

Bedbugs, an ancient scourge long thought eradicated, have again become a serious problem in the USA, creating misery, financial hardship, and health hazards (Schulz 2011). Government agencies associate the alarming resurgence of bedbugs with significant public health importance. Despite actions taken to date by governmental bodies, investigators in various relevant areas claim that federal agencies are failing in their duties (Schulz 2011) to adequately deal with this resurgent problem.

P. Kovacic (✉)
Department of Chemistry and Biochemistry, San Diego State University,
San Diego, CA 92182, USA
e-mail: pkovacic@sundown.sdsu.edu

R. Somanathan
Centro de Graduados e Investigación del Instituto Tecnológico de Tijuana,
Apdo postal, 1166 Tijuana, BC, Mexico

D.M. Whitacre (ed.), *Reviews of Environmental Contamination and Toxicology*,
Reviews of Environmental Contamination and Toxicology 218,
DOI 10.1007/978-1-4614-3137-4_4, © Springer Science+Business Media, LLC 2012

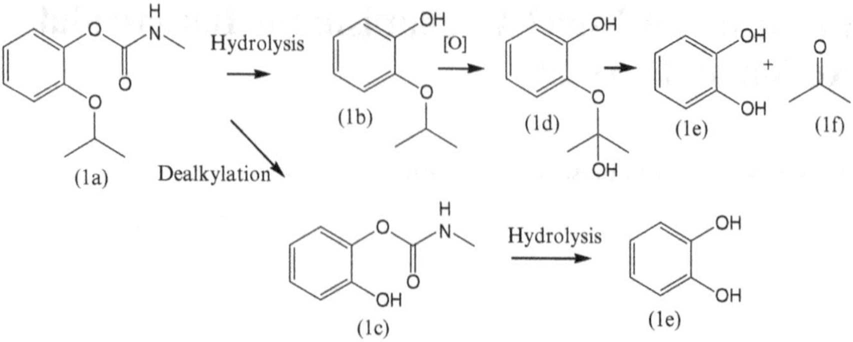

Fig. 1 (a–f) A scheme showing the metabolism of propoxur (1a). This carbamate hydrolyzes to form isopropoxy phenol (1b) and/or may undergo ether dealkylation to generate o-hydroxy phenyl methylcarbamate (1c). The metabolite (1b) further degrades via oxidation to an unstable hemiacetal (1d), which rapidly decomposes to (1e) and acetone (1f). Hydrolysis of (1c) also yields catechol (1e)

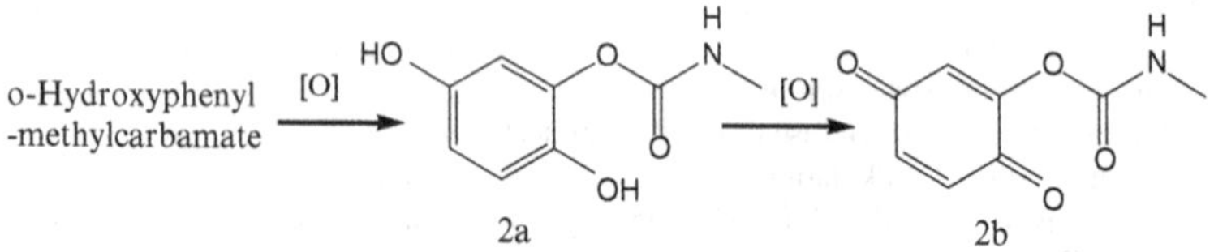

Fig. 2 Ring hydroxylation of o-hydroxy phenyl methylcarbamate to form hydroquinone methylcarbamate (2a), which conceivably is oxidized to the proposed benzoquinone methylcarbamate (2b)

Bedbugs show an increased resistance to older insecticides and only a limited number of potentially efficacious chemicals are available to help mitigate the current outbreak. There have been calls by industry for the U.S. EPA to grant an emergency exemption of the carbamate insecticide propoxur (Fig. 1a); propoxur is one of the few treatment options available that provides effective primary and residual kill of bedbugs. However, thus far, EPA has not deigned to allow use of propoxur, primarily because of concerns for its public health hazard.

The generally accepted mechanism of propoxur's toxic action entails inhibition of cholinesterase (ChE), as is the case for most carbamate insecticides. However, there is a considerable body of research that supports the view that many physiologically active substances act at multifaceted sites, and this may also be the case for propoxur. In this review, we provide lines of evidence to support the view that electron transfer (ET)–reactive oxygen species (ROS)–oxidative stress (OS) play a role in both the insecticidal action of propoxur and the adverse human effects it causes.

Propoxur is a catechol derivative that contains carbamate and isopropyl groups on the oxygen. Metabolic studies (Figs. 1 and 2) have revealed that hydrolysis of the carbamate and dealkylation of the isopropyl occurs to yield the parent catechol. Further nuclear hydroxylation yields a hydroquinone derivative. Both of these

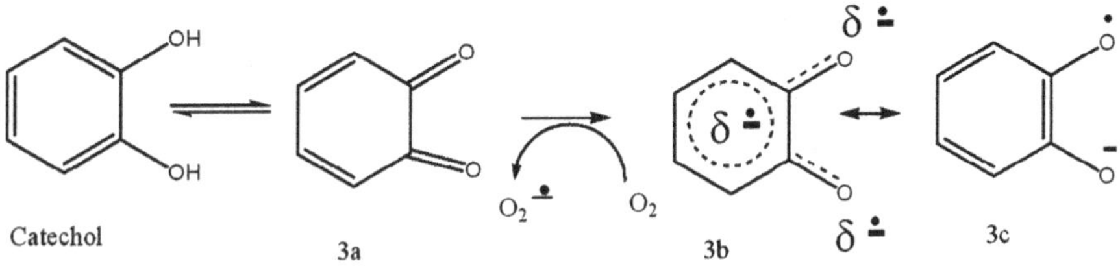

Fig. 3 (a–c) This scheme depicts redox cycling of catechol and the proposed *o*-quinone (3a), including the ET of 3a via the mechanism shown (3b) to form the radical superoxide anion (3b and 3c), with the formation of superoxide

metabolic products are able to undergo redox cycling with the corresponding quinone to produce ROS. ROS are known to selectively participate in pesticide action and in human toxicity.

Prior literature, cited below, demonstrates that ET–ROS–OS play a role in the toxic action of many drugs and toxins. A preponderance of bioactive substances or their metabolites incorporate ET functionalities, which, we believe, play an important role in physiological responses, examples of which include the following: quinones (or phenolic precursors), metal complexes (or complexors), nitroso and hydroxylamine metabolites from $ArNO_2$ and $ArNH_2$ (Kovacic and Somanathan 2011), and conjugate diimines (or iminium species). In vivo redox cycling of such entities with oxygen can occur, giving rise to OS through generation of ROS (e.g., hydrogen peroxide, hydroperoxides, alkylperoxides, and diverse radicals such as hydroxyl, alkoxyl, hydroperoxyl, and superoxide), as illustrated in Fig. 3a–c. In some cases, ET results in interference with normal electrical effects, such as that occurs in respiration or neurochemistry. Generally, active entities possessing ET groups display reduction potentials in the physiologically responsive range, i.e., more positive than –0.5 V. ET, ROS, and OS have been increasingly implicated in the mode of action of many drugs and toxicants, such as the following: anti-infective agents (Kovacic and Becvar 2000), anticancer drugs (Kovacic and Osuna 2000; Kovacic 2007), carcinogens (Kovacic and Jacintho 2001a), reproductive toxins (Kovacic and Jacintho 2001b), nephrotoxins (Kovacic et al. 2002), hepatotoxins (Poli et al. 1989), cardiovascular toxins (Kovacic and Thurn 2005), nerve toxins (Kovacic and Somanathan 2005), mitochondrial toxins (Kovacic et al. 2005), abused drugs (Kovacic and Cooksy 2005), immunotoxins (Kovacic and Somanathan 2003), pulmonary toxins (Kovacic and Somanathan 2009), dermal toxins (Kovacic and Somanathan 2010), ototoxins (Kovacic and Somanathan 2008a), eye toxins (Kovacic and Somanathan 2008b), thyroid toxins (Kovacic and Edwards 2010), and also in other categories, such as human illnesses (Halliwell and Gutteridge 1999). The phenolic class, a focus of this report, is structurally represented in therapeutics, carcinogens, and toxicants. Of particular interest is catechol, which can redox cycle with *o*-quinone.

There is a plethora of experimental evidence, as indicated above, that supports the theoretical framework for OS as being involved in the mode of action of those chemicals mentioned above. They act by mechanisms that include generating common ROS, lipid peroxidation, by producing degradation products of oxidation, depleting AOs, affecting exogenous AOs, oxidizing DNA to produce cleavage products. This mechanism is comprehensive and unifying, and is in keeping with the observation that many ET substances display a variety of activities, e.g., multiple drug properties, as well as toxic effects.

2 Metabolism of Propoxur: Support for an ET-ROS-OS Mechanism

Extensive research has been performed on propoxur metabolism, which lends credence to the mechanistic framework described above. Most of these metabolic data are outlined in a 1989 report (Pesticide Residues 1987). Two initial propoxur metabolic routes involve hydrolysis of this carbamate to form isopropoxy phenol (Fig. 1b), and ether dealkylation to generate o-hydroxy phenyl methylcarbamate (Fig. 1c). Both products then serve as precursors of catechol (Fig. 1e). Loss of the isopropyl group apparently occurs by oxidation to form an unstable hemiacetal (Fig. 1d), which further decomposes to catechol (Fig. 1e) and acetone (Fig. 1f) (Shrivastava et al. 1970). Similar dealkylation of phenolic methyl ethers is well known (Kovacic and Somanathan 2004; Kovacic and Osuna 2000). The parent propoxur, as well as its metabolites, may undergo ring hydroxylation at various positions (Chemical Watch Factsheet 2010, 2011). One of the hydroxylation products of particular interest is hydroquinone methylcarbamate (Fig. 2a). The metabolites of most importance are catechol and the hydroquinone methylcarbamate (Fig. 2a), both of which may participate in redox cycling with the corresponding proposed quinones, namely o-quinone (Fig. 3a) and the proposed p-benzoquinone methylcarbamate (Fig. 2b).

Quinones, especially the *ortho* isomers, display favorable reduction potentials, facilitate redox cycling in vivo, and form ROS. Quinone metabolites are sometimes difficult to isolate, because they are easily attacked by protein nucleophiles, such as side-chain amino and thiol groups of amino acids. For the proposed quinone metabolite of propoxur, the resulting ROS may be responsible, at least in part, for this compounds insecticidal action and toxicity. Site binding to proteins of any such proposed quinone metabolite would affect toxic action. There is extensive support for such a scheme that involves ET–ROS–OS, as indicated in the Introduction.

Related research papers that have also addressed the metabolism of propoxur include the following: Keiser et al. 1983; Suma et al. 2005; Brouwer et al. 1993; Abd-Elraof et al. 1981.

3 ROS and OS

The literature that supports the role that ROS and OS has in the action of therapeutics, carcinogens, and toxicants is extensive, and the involvement of ROS–OS in the physiological action of propoxur has been documented in many research reports.

For example, low dose exposure to propoxur induced an increase in lipid peroxidation, and this increase was much more significant at higher doses (Makhija and Pawar 1975). Malonaldehyde was also identified as being present. This is significant, because the presence of malonaldehyde indicates that in vivo lipid peroxidation had occurred. Banerjee et al. (1999) studied OS in blood obtained from propoxur poisoning cases and found an increase in lipid peroxidation, coupled with altered levels of GSH (glutathione) and ROS scavenging enzymes. In another study involving lipid peroxidation, results suggest that this carbamate could injure cells via OS in ways that involved malondialdehyde production (Maran et al. 2010), with GSH bestowing protection against cell damage. Simultaneous treatment with melatonin markedly attenuated the effect propoxur had on lipid peroxidation, OS, and immunotoxicity (Suke et al. 2006). The results of the preceding studies are augmented by additional research performed on lipid peroxidation, AO enzymes, and the GSH redox systems (Seth et al. 2000; Makhija and Pawar 1978; Yadav et al. 2010; Mehta et al. 2010).

Because pesticide use is widespread, genotoxic effects from exposure to them are of concern (Undeger and Basaran 2005). Various authors have performed studies that provide evidence that insecticides, such as propoxur, possess genotoxic properties. Propoxur significantly increased DNA damage in human lymphocytes. Such an adverse effect is often associated with attacks by ROS. Exposure to carbamates, including propoxur, resulted in GSH depletion (Maran et al. 2009), which enhanced GSH transferase activity that serves as a countermeasure to the ROS–OS.

4 Toxicity of Propoxur

Propoxur is reported to be highly toxic (Chemical Watch Factsheet 2010, 2011). Early poisoning symptoms include malaise, dizziness, muscle weakness, sweating, headache, nausea, and diarrhea. Other adverse effects caused by propoxur are anemia, leukemia, tumor formation, bladder hyperplasia, and sciatic neuropathy. Studies with rats also revealed decreased motor activity (Syrowatka et al. 1971). Reports on propoxur's toxic effects have also addressed its acute toxicity on carbohydrate metabolism (Srivastava and Singh 1982), and chronic toxic effects on carbohydrate, protein, and serum electrolyte levels (Singh et al. 1997). Modulations in ionic composition were seen in the rat brain from propoxur treatment, indicating an impairment of electric neuronal activity, oxygen consumption, the ATPase system, and disruption in ion movement (Babu et al. 1990).

In addition to the genotoxic effects ascribed to propoxur (see Sect. 3), considerable attention has been devoted to the carcinogenic effects of this insecticide as well. Propoxur is known to produce tumors in rats (Cohen et al. 1994). Moreover, tumor generation and leukemia induction that were associated with propoxur exposure have been noted in the literature (Pesticide Residue 1987). Prenatal exposure to the insecticide may predispose to leukemia-associated chromosomal translocations (Lafiura et al. 2007). Use of propoxur as a household insecticide was associated with acute leukemia in children (Kuo et al. 2008). Propoxur is also notorious for its propensity to transform into a highly genotoxic N-nitroso derivative, which has been implicated in increased potential for genetic instability and possibly the enhancement of malignancy potential of treated cells (Kuo et al. 2008). In another study that addressed the genotoxicity of propoxur and its N-nitroso derivative, the derivative was found to be more cytotoxic than the parent compound (Wang et al. 1998). Propoxur inhibited gap-junctional intercellular communication, indicating action through some epigenetic mechanisms, such as tumor promotion or cell proliferation, in the multiple process of chemical carcinogenesis.

5 Mechanisms of Propoxur's Action

5.1 Inhibition of Cholinesterase or Acetylcholinesterase

The inhibition of ChE or acetylcholinesterase (AChE) is generally accepted as the principal way in which the carbamate insecticides kill organisms (Xiao et al. 2010). Behavioral changes that occur during intoxication with low doses of propoxur may accompany brain ChE activity inhibition (Thiesen et al. 1999). Results suggest that repeated inhibition of AChE activity may cause downregulation of muscarinic AChE receptors in the hippocampus, when the insecticide is injected (Kobayashi et al. 2007). A decrease in motor activity is highly predictive of ChE inhibition of N-methylcarbamates (McDaniel et al. 2007). Data suggest that disruption of cortical processing of visual signs was related to inhibition of brain ChE activity by N-methylcarbamates, including propoxur (Mwanza et al. 2008). Exposure to ChE-inhibitor insecticides has been associated with many serious toxicities due to the accumulation of acetylcholine in various body sites (Moody and Terp 1988). Erythrocyte ChE levels showed a significant decrease, related to the increasing length of exposure to propoxur (Weisbroth et al. 1983).

5.2 Aromatic Ether Dealkylation

The dealkylation of aromatic ethers, as illustrated in this chapter by the metabolic loss of the isopropyl group from propoxur, is widely reported in the literature, especially for the methyl group, e.g., mescaline and etoposide (Kovacic and Somanathan 2009; Kovacic and Osuna 2000). When aromatic ether dealkylation occurs, a hemiacetal metabolite formed by oxidation appears to be an intermediate, since the end products are a phenol derivative and formaldehyde.

5.3 Mechanisms Other than ET–ROS–OS

Other data suggest that the carbamate insecticides, such as propoxur, may represent a class that induces toxicity through a mechanism other than from ligand binding, and therefore may act as general endocrine modulators in mammalian cells

(Klotz et al. 1997). It has been suggested that the mechanism of lethality caused by ChE inhibitors could be interpreted as having created a balance between anti-ChE activity and some other mechanism (Takahashi et al. 1994).

Few studies have been performed in which ET–ROS–OS have been investigated as causative agents in causing toxic action. However, judging from the aforegoing discussion, it is reasonable to propose that a multipronged mechanistic attack is possible for propoxur, as has occurred for other physiologically active agents.

5.4 Other Pesticides

Propoxur's induction of toxicity through a mechanism involving ROS–OS obtains credibility from reports of similar action that involve other pesticides. Considerable research points to participation of ROS–OS in neurotoxicity induced by certain pesticides, including metal compounds, synthetic pyrethroids, chlorinated hydrocarbons (aldrin, dieldrin, endrin, and hexachlorocyclohexane), 2,4-dinitrophenol, and certain organophosphates (Kovacic and Somanathan 2005; Kovacic 2003). Numerous routes of formation are described for these pesticides. Dinitrophenol is a close relative to ET agents, since it possesses both a phenolic group and nitro groups, both of which are precursors of redox couples.

6 Summary

Propoxur is a carbamate insecticide that has recently attracted considerable attention as a possible treatment option for addressing the bedbug epidemic. The generally accepted mechanism of toxicity for propoxur involves the inhibition of ChE, as is the case for many agents in the category. Considerable research supports the concept that most physiologically active substances induce their effects through multifaceted action. In this review, we provide evidence that ET–ROS–OS participate mechanistically in both the action and in human toxicity of pesticides, including propoxur. Propoxur is a catechol derivative that contains carbamate and isopropyl groups on the oxygens in its moiety. Metabolic studies with propoxur reveal hydrolysis of the carbamate and dealkylation of the isopropyl group to yield the parent catechol. In addition, nuclear hydroxylation produces a hydroquinone derivative. Both the catechol and this hydroquinone derivative are potentially able to undergo redox cycling with the corresponding quinone to produce ROS.

It is primarily for these reasons that we believe propoxur may be similar to other classes of physiologically active compounds in producing effects through ET–ROS–OS. Generally, reactive ROS are generated by metabolic processes that yield ET entities, and this occurs with propoxur as well. Although ROS are commonly associated with toxicity, there is little recognition in the literature that they can also play a role in therapeutic action.

Acknowledgments We are grateful to Ashley Berry and Thelma Chavez for editorial assistance.

References

Abd-Elraof TK, Dauterman WC, Mailman RB (1981) In vivo metabolism and excretion of propoxur and malathion in the rat: effect of lead treatment. Toxicol Appl Pharmacol 59:324–330

Babu GR, Reddy GR, Reddy AT, Rajendra W, Chetty CS (1990) Modulations in ionic composition and ATPase system in the brain of albino rat under induced propoxur toxicity. Biochem Int 21:1105–1111

Banerjee BD, Seth V, Bhattacharya A, Pasha ST, Chakraborty AK (1999) Biochemical effects of some pesticides on lipid peroxidation and free-radical scavengers. Toxicol Lett 107:33–47

Brouwer R, Van Maarleveld K, Ravensberg L, Meuling W, de Kort W, van Hennen JJ (1993) Skin contamination, airborne concentrations, and urinary metabolite excretion of propoxur during harvesting of flowers in greenhouses. Am J Ind Med 24:593–603

Chemical watch factsheet (2010–2011), Propoxur. Pesticides and You. 30(4): Winter 2010–11

Cohen SM, Cano M, Johnson LS, St. John MK, Asamoto M, Garland EM, Thyssen JH, Sangha GK, Van Goethem DL (1994) Mitogenic effects of propoxur on male rat bladder urothelium. Carcinogenesis 15:2593–2597

Halliwell B, Gutteridge JMC (1999) Free radicals in biology and medicine. Oxford University Press, New York, pp 1–897

Keiser JE, Kirby KW, Trmmel F (1983) Reversed-phase high-performance liquid chromatography separation of human phenolic metabolites of propoxur (Baygon), carbofuran and carbaryl. J Chromatogr 259:186–188

Klotz DM, Arnold SF, McLachlan JA (1997) Inhibition of 17 beta-estradiol and progestrone activity in human breast and endometrial cancer cells by carbamate insecticides. Life Sci 60:1467–1475

Kobayashi H, Suzuki T, Sakamoto M, Hashimoto W, Kashiwada K, Sato I, Akahori F, Satoh T (2007) Brain regional acetylcholinesterase activity and muscarinic acetylcholine receptors in rats after repeated administration of cholinesterase inhibitors and its withdrawal. Toxicol Appl Pharmacol 219:151–161

Kovacic P (2003) Mechanism of organophosphates (nerve gases and pesticides) and antidotes: electron transfer and oxidative stress. Curr Med Chem 10:2705–2709

Kovacic P (2007) Unifying mechanism for anticancer agents involving electron transfer and oxidative stress: clinical implications. Med Hypotheses 69:510–516

Kovacic P, Becvar LE (2000) Mode of action of anti-infective agents: emphasis on oxidative stress and electron transfer. Curr Pharm Des 6:143–167

Kovacic P, Cooksy AL (2005) Unifying mechanism for toxicity and addiction by abused drugs: electron transfer and reactive oxygen species. Med Hypotheses 64:366–367

Kovacic P, Edwards C (2010) Integrated approach to the mechanisms of thyroid toxins: electron transfer, reactive oxygen species, oxidative stress, cell signaling, receptors, and antioxidants. J Recept Signal Transduct Res 30:133–142

Kovacic P, Jacintho JD (2001a) Mechanisms of carcinogenesis: focus on oxidative stress and electron transfer. Curr Med Chem 8:773–796

Kovacic P, Jacintho JD (2001b) Reproductive toxins: pervasive theme of oxidative stress and electron transfer. Curr Med Chem 8:863–892

Kovacic P, Osuna JA (2000) Mechanisms of anticancer agents: emphasis on oxidative stress and electron transfer. Curr Pharm Des 6:277–309

Kovacic P, Somanathan R (2003) Integrated approach to immunotoxicity: electron transfer, reactive oxygen species, antioxidants, cell signaling and receptors. J Recept Signal Transduct Res 28:323–346

Kovacic P, Somanathan R (2004) Novel, unifying mechanism for mescaline in the central nervous system. Oxid Med Cell Longev 2:181–190

Kovacic P, Somanathan R (2005) Neurotoxicity: the broad framework of electron transfer, oxidative stress and protection by antioxidants. Curr Med Chem-CNS Agents 5:249–258

Kovacic P, Somanathan R (2008a) Ototoxicity and noise trauma; electron transfer, reactive oxygen species, cell signaling, electrical effects, and protection by antioxidants; practical medical aspects. Med Hypotheses 70:914–923

Kovacic P, Somanathan R (2008b) Unifying mechanism for eye toxicity: electron transfer, reactive oxygen species, antioxidant benefits, cell signaling and cell membranes. Cell Membr Free Radic Res 1:56–69

Kovacic P, Somanathan R (2009) Pulmonary toxicity and environmental contamination: radicals, electron transfer, and protection by antioxidants. In: Whitacre DM (ed) Rev Environ Contam Toxicol, Springer, New York 201:41–69

Kovacic P, Somanathan R (2010) Dermal toxicity and environmental contamination: electron transfer, reactive oxygen species, oxidative stress, cell signaling, and protection by antioxidants. In: Whitacre DM (ed) Rev Environ Contam Toxicol. Springer, New York, 203:119–138

Kovacic P, Somanathan R (2011) Novel unifying mechanism for aromatic primary-amines (therapeutics, carcinogens and toxicants): electron transfer, reactive oxygen species, oxidative stress and metabolites. Med Chem Commun 2:106–112

Kovacic P, Thurn LA (2005) Cardiovascular toxins from the perspective of oxidative stress and electron transfer. Curr Vasc Pharmacol 3:107–117

Kovacic P, Sacman A, Wu-Weis M (2002) Nephrotoxins: widespread role of oxidative stress and electron transfer. Curr Med Chem 9:823–847

Kovacic P, Pozos RS, Somanathan R, Shangari R, O'Brien PJ (2005) Mechanism of mitochondrial uncouplers, inhibitors, and toxins: focus on electron transfer, free radicals, and structure-activity relationships. Curr Med Chem 5:2601–2623

Kuo HH, Shyu SS, Wang TC (2008) Genotoxicity of low dose N-nitroso propoxur to human gastric cells. Food Chem Toxicol 46:1619–1626

Lafiura KM, Bielawski DM, Posecion NC Jr, Ostrea EM Jr, Matherly LH, Taub JW, Ge Y (2007) Association between prenatal pesticide exposures and the generation of leukemia-associated T (8;21). Pediatr Blood Cancer 49:624–628

Makhija SJ, Pawar SS (1975) Alterations in hepatic drug metabolism and lipid peroxidation during administration of Baygon, a pesticide. Chem Biol Interact 10:295–299

Makhija SJ, Pawar SS (1978) NADH dependent aminopyrine N-demethylation and lipid peroxidation during Baygon intoxication in hepatic and extrahepatic guinea pig tissues. Bull Environ Contam Toxicol 19:538–544

Maran E, Fernández-Franzón M, Barbieri P, Font G, Ruiz MJ (2009) Effects of four carbamate compounds on antioxidant parameters. Ecotoxicol Environ Saf 72:922–930

Maran E, Fernández-Franzón M, Font G, Ruiz MJ (2010) Effects of aldicarb and propoxur on cytotoxicity and lipid peroxidation in CHO-K1 cells. Food Chem Toxicol 48:1592–1596

McDaniel KL, Padilla S, Marshall RS, Phillips PM, Podhorniak L, Qian Y, Moser VC (2007) Comparison of acute neurobehavioral and cholinesterase inhibitory effects of N-methylcarbamates in rat. Toxicol Sci 98:552–560

Mehta KD, Garg GR, Mehta AK, Arora T, Sharma AK, Khanna N, Tripathi AK, Sharma KK (2010) Reversal of propoxur-induced impairment of memory and oxidative stress by 4′-chlorodiaepam in rats. Naunyn Schmiedebergs Arch Pharmacol 381:1–10

Moody SB, Terp DK (1988) Dystonic reaction possibly induced by cholinesterase inhibitor insecticides. Drug Intell Clin Pharm 1(22):311–312

Mwanza JC, Finley D, Spivey CL, Graff JE, Herr DW (2008) Depression of the photic after discharge of flash evoked potentials by physostigmine, carbaryl and propoxur, and the relationship to inhibition of brain cholinesterase. Neurotoxicology 29:87–100

Pesticide residues in food-1987. Part II-Toxicology, Food Agriculture Organization of the United Nations. pp 186, 187

Poli G, Cheeseman KH, Dianzani MU, Slater TF (1989) Free radicals in the pathogenesis of liver injury. Pergamon, New York, pp 1–330

Schulz WG (2011) Battling the bedbug epidemic. Chem Eng News 89:13–18

Seth V, Banerjee BD, Bhattacharya A, Chakravorty AK (2000) Lipid peroxidation, antioxidant enzymes, and glutathione redox system in blood of human poisoning with propoxur. Clin Biochem 33:683–685

Shrivastava SP, Georghiou GP, Metcalf RL, Fukuto TR (1970) Carbamate resistance in mosquitoes. The metabolism of propoxur by susceptible and resistant larvae of Culex pipiens fatigans. Bull World Health Organ 42:931–942

Singh NN, Das VK, Srivastava AK (1997) Chronic toxicity of propoxur on carbohydrate, protein and serum electrolyte levels in catfish, Heteropneustes fossilis. Biomed Environ Sci 10:408–414

Srivastava AK, Singh NN (1982) Acute toxicity of propoxur on carbohydrate metabolism of Indian catfish (Heteropneustes fossilis). Toxicol Lett 11:31–34

Suke SG, Kumar A, Ahmed RS, Chakraborti A, Tripathi AK, Mediratta PK, Banerjee BD (2006) Protective effect of melatonin against propoxur suppression of humoral immune response in rats. Indian J Exp Biol 44:312–314

Suma R, Sarin RK, Saiprakash PK, Ramakrishna S (2005) Simple liquid chromatographic method for the rapid and simultaneous determination of propoxur and its major metabolic isopropoxy phenol in rat blood and urine using solid-phase extraction. J Anal Toxicol 29:728–733

Syrowatka T, Jurek A, Nazarewicz T (1971) Short-term chronic study on the toxicity of o-isopropoxypheny-N-methylcarbamate (propoxur) in rats. Rocz Panstw Zakl Hig 22:579–589

Takahashi H, Kakinuma Y, Futagawa H (1994) Non-cholinergic lethality following intravenous injection of carbamate insecticide in rabbits. Toxicology 93:195–207

Thiesen FV, Barros HM, Tannhauser M, Tannhauser SL (1999) Behavioral changes and chlinesterase activity of rats acutely treated with propoxur. Jpn J Pharmacol 79:25–31

Underger U, Basaran N (2005) Effects of pesticides on human peripheral lymphocytes in vitro: induction of DNA damage. Arch Toxicol 79:169–176

Wang TC, Chiou JM, Chang YL, Hu MC (1998) Genotoxicity of propoxur and its N-nitroso derivative in mammalian cells. Carcinogenesis 19:623–629

Weisbroth SP, Weisbroth SH, Grey RM (1983) Effect of propoxur-impregnated pesticide tape on mouse cholinesterase levels. Lab Anim Sci 33:151–153

Xiao LS, Dou W, Li Y, Wang JJ (2010) Comparative studies of acetylcholinesterase purified from three field populations of Liposcelis entomophila (enderlein) (psocoptera: liposcelididae). Arch Insect Biochem Physiol 75:158–173

Yadav CS, Kumar V, Suke SG, Ahmed RS, Mediratta PK, Banerjee BD (2010) Propoxur-induced acetycholine esterase inhibition and impairment of cognitive function: attenuation by *Withania somnifera*. Indian J Biochem Biophys 47:117–120

Index